Show Me the Fair:
A HISTORY OF THE
MISSOURI STATE FAIR

By Rhonda Chalfant

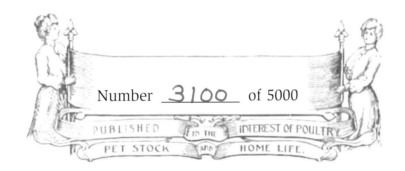

Number 3100 of 5000

PUBLISHED IN THE INTEREST OF POULTRY
PET STOCK AND HOME LIFE.

THE
DONNING COMPANY
PUBLISHERS

Show Me the Fair:

A HISTORY OF THE MISSOURI STATE FAIR

By Rhonda Chalfant

The Donning Company Publishers
184 Business Park Drive, Suite 206
Virginia Beach, VA 23462

Walsworth Publishing Company
306 N. Kansas Avenue
Marceline, MO 64658

Steve Mull, General Manager
Ed Williams, Project Director
Dawn V. Kofroth, Assistant General Manager
Sally Clarke Davis, Editor
Elizabeth McClure, Graphic Designer
Scott Rule, Director of Marketing

ISBN 1-57864-189-6

Printed in the United States of America

CONTENTS

PREFACE

The task of documenting one hundred years of a state fair is fraught with peril. An incredible amount of primary source material is available in official papers found in archives and special collections; another sort of information is readily available in newspaper accounts. Postcards, trade cards, tickets, and other paper items add to the amount of source material, as do souvenirs collected over the course of one hundred years. Almost everyone who has been to the fair has some memory to share. In addition, an immense amount of secondary source material is available. The agricultural fair, forming as it does an integral part of our culture, has been described and analyzed repeatedly. The decisions involved in choosing what to include and what to omit in such a short work when so much information is available create, sometimes, nightmares for the writer.

This work represents an attempt to place the Missouri State Fair in the context of its time. The dates of the fair, from its proposal in 1899 to the one hundredth fair in 2002, have encompassed rapid and startling changes in nearly all aspects of our culture. The fair has, naturally, changed as the world has changed. This history seeks to explain the relationship between the changes in agriculture, transportation, science, technology, the arts, and lifestyle and the changes in the fair itself. This work attempts to reflect the richness of the fair's history through words and through a small sampling of the thousands of photos and collectibles that capture some of the highlights of Missouri's showcase of the best of the best.

First Missouri State Fair Commission

From left to right: John Saunders, Brenda Lampton, Missouri State Fair Director Gary Slater, Betty Linke, Larry Foster, Norwood Creason, Jerry Divin, Lowell Mohler, Sean McGinnis, and Ida Cox

2002 Missouri State Fair Commission

Front row: Larry Foster, Kent Blades, Betty Linke. *Back row:* Missouri State Fair Director Mel Willard, Norwood Creason, Morris Brown, Ida Cox, Lowell Mohler, Jerry Divin, Sean McGinnis

ACKNOWLEDGMENTS

Many people helped with the production of this work by finding information, by making collections available, by explaining technology, by gathering photographs, and by reading and editing the manuscript. Thanks are due to all of them.

First, without Van Beydler, this book would not exist. His collection of State Fair memorabilia and his expertise with scanners, photography, and e-mail contributed greatly to the illustrations and the text. Others, including Tony Perryman, Doug Cline, R. Terrell Wright, Sidney Brink of the *Sedalia Democrat*, Charles Wise, Sue Brockman, A. J. Heck, and Caroline Thomas provided photographs, both new and old. Doug, Charles, and Tony also provided souvenirs to be photographed.

The State Historical Society of Missouri's newspaper collection provided a valuable source of information, as did their photo collection. The Missouri State Archives staff, particularly photography curator Laura Jolly, were gracious and helpful. The Archives was the recipient, several years ago, of a large collection of papers and photos from the state fair. Their staff catalogued and indexed the collection, a task rendered very difficult by the absence of labels on most of the photos. This collection was invaluable. The Sedalia Public Library, Boonslick Regional Library, and State Fair Community College Library provided books, and Sedalia Public Library allowed items from their files to be scanned and included. In addition, the Sedalia Public Library provided microfilm copies of the Sedalia papers. The *Sedalia Democrat* allowed photographs and information from a book they published to be used. In the interest of brevity, photos are credited with the initials of their producers or owners, who have been named above. In an attempt to keep the text easy to read, the committee decided against the use of footnotes or endnotes. Sources of quotations are identified in the text and in the Works Consulted section at the end of the volume.

Thanks are also due to those who helped by providing information, especially Diane Cole, Ray Hagan, Ida Shobe, Mike Swain, Rick Yeager, Dave Goodson, Maxine Griggs, Faith Lovell, Myra Roberts Finks, Howard Roberts, Caroline Thomas, A. J. Heck, Sandra Blunk, and Madge Gressley, and members of the Pettis County Historical Society.

Members of the Centennial Committee, many of who work for state agencies, provided information about their work, as did department superintendents. Thanks are due to Commissioners Betty Linke, Ida Cox, Morris Brown, and Norwood Creason and committee members Tony Perryman, Tracy Ritter, Jim Shoemaker, David Dick, Greg Wood, Jim McCarty, Matt Boatright, Jim Mathewson, Robert Ramey, and Deb Biermann. The fair staff— Mel Willard, Kim Allen, Wendy Baker, Leslie Brantley, Melba MacDonald, Nancy Bouse, Vickie Embree, Jo Colvin, Mike Riley, and Robert Howell—consistently gave support, encouragement, and information.

Thanks to everyone who helped. Any errors are solely my responsibility, and all praise is theirs.

Rhonda Chalfant

This book was made possible

in part by support from

Central Bank of Missouri, Sedalia, Missouri

and

Third National Bank, Sedalia, Missouri

One Hundred Fairs of Fun: Continuity and Change

When Missouri Governor Lon Stephens signed House Bill 279 on April 19, 1899, he set in motion a series of events he probably couldn't imagine. One hundred fairs later, the Missouri State Fair has grown from a five-day event with attendance of 25,346 to an eleven-day extravaganza regularly drawing nearly 350,000 visitors.

Fair aficionado Noel Perrin identifies a point at which fair organizers had

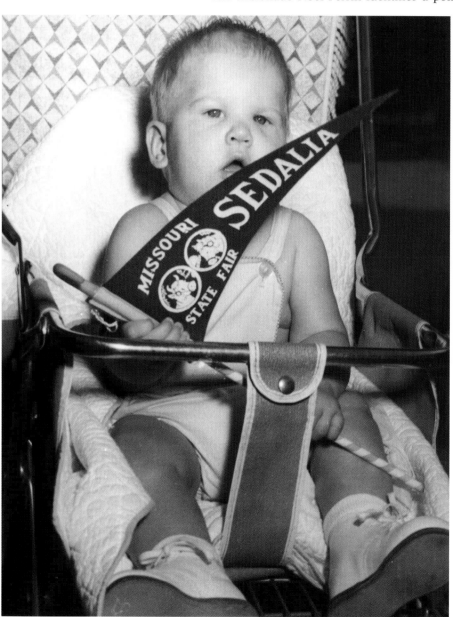

to add to the educational nature of the state fair to draw visitors; the "frivolity" fair managers added Perrin identifies as "politicians and fast horses." The Missouri State Fair has seen plenty of both. Presidents Taft, Truman, and Reagan have visited the fair. Dan Patch, billed as "the world's fastest horse," demonstrated his speed in 1909. Racehorses, a primary feature of the first fair, still run the half-mile track.

The fair is, in a sense, a microcosm of the world. It reflects, in 396 acres and eleven days, the events of the world. World Wars I and II, the Cold War, and space exploration made their presence known at the fair, as did the nation's Bicentennial Celebration in 1976. Changes in music and popular entertainment reveal themselves in the shift from vaudeville to variety shows to concerts by individual stars. Clothing

worn by visitors reflects changes in attitudes as well as in styles. The corseted, hatted, and gloved ladies in proper shirtwaists accompanying gentlemen in vests and jackets at the first fair slowly gave way to visitors wearing the more comfortable shorts and t-shirts of today.

At the same time, the fair is uniquely Missouri. *The Premium List, Missouri State Fair, 1901,* claimed, "The Missouri State Fair will be what Missourians may make it. It should be the greatest institution of the kind on earth" because of "the fertility of our soil, the great variety of our products, extended deposits of valuable minerals, large areas of valuable forests, the superiority of our herds, and the progressive character of our citizenship." The fair continues to showcase the best of Missouri agriculture, art, and products.

The fair shows the world as it is, with its numbers of people, diversity of visitors, and varieties of exhibits. Young families, their toddlers in strollers, mingle with senior citizens. Tattooed men and women display their bodies as art while they socialize with cowboys and carnies. Teens congregate along the Midway. Children talk to the clowns, pet animals, cheer their favorite at the pig races, and ride the carousel. Yet the fair also presents an opportunity to escape everyday realities. The lights of the carnival create a neon wonderland. The upside-down twists of a Midway ride provide a new view of the world. In the camaraderie of the crowd at a concert, everyone is a friend.

The fair is notable for its continuity—the things about the fair that remain unchanged—especially considering the immense changes in agriculture, technology, and popular culture since 1901. Some events have remained strikingly similar. The first fair, for example, drew crowds to watch chicks hatch in a

The Wright Brothers bring their airplane to the fair on September 29, 1910. (R. T. W.)

Facing page: A young fairgoer enjoys a souvenir pennant. (M. S. A.)

The Midway featured many rides in 1941. (V. B.)

Below: Machinery exhibits have always shown the latest in farm equipment. (M. S. A.)

Sure-Hatch incubator; the baby chicks at the Children's Barnyard are still a huge hit with both adults and children alike. Furniture stores displayed furniture suites in model rooms at early fairs; today manufactured homes show the latest in building techniques and convenient living. In the early years of the fair, self-styled critics denounced the judges' choice of prize-winning paintings. Visitors still admire some art exhibits and puzzle over how other pieces could have won prizes. The *Official Program* from the first fair announced the "Greatest Livestock Exhibit on Earth"; *Successful Farming* identifies the Missouri State Fair today as having the "most livestock of any state fair."

But even in the continuity there is change. The Wright Brothers flying

machine was a hit in 1909; the B-2 Stealth Bomber now does flyovers. The Elks Street Fair, a downtown event accompanying the first fair, featured a Midway of shows and games; now the Midway with its games, rides, and shows, is part of the fair.

The Missouri State Fair began as an educational institution, a place where Missouri residents could learn better methods of farming, of stock breeding, of homemaking, and where they could see the latest innovations that could make their lives easier. In 1925, the state fair issued a press release stating, "The Missouri State Fair fulfills its chief purpose and aim in placing agricultural products as one of the main features of the Fair. The prime reason for any fair's existence is education—education for every age and class of citizen within its boundaries."

Two well-dressed couples had their photograph taken at the 1909 fair. (D. C.)

Left: The *Missouri Ruralist* offered a souvenir button to extend a welcome to fair-goers in the 1920s. (V. B.)

The fair still fulfills its purpose of education. Missourians are able to see the best of Missouri's agricultural and industrial products, learn the dangers of train accidents, or trace the development of a fetus. Cooks can see the latest in kitchen technology demonstrated at the Home Economics building. Visitors can see Missouri wildlife at the conservation department exhibit; children can listen to Smokey the Bear or talk to Otto the Talking Car.

The fair is also a commercial enterprise. At the first fair, Missouri dairymen could buy the newest in cream separators. In the early part of the century, automobile companies displayed and sold cars; machinery manufacturers showed plows and windmills. Missourians now have the chance to buy the newest vegetable choppers, non-stick pots and pans, vibrating chairs, quilting machines, wood-burning furnaces, log splitters, and portable corral panels.

Visitors also buy souvenirs. At the first fair, Bichsel's Jewelry Company offered commemorative spoons; other dealers offered napkin rings, pins, and dolls. Now, souvenir hunters bag foam alligators that appear to walk at the end of a wire leash, sport airbrushed t-shirts, and treasure heart-shaped pendants engraved with their beloved's name. Those not willing to spend hard-earned dollars on souvenirs can collect yardsticks, cardboard fans, tote bags, and enough recipes to make up a fair sized cookbook. They can sample Missouri grape juice, summer sausage, honey, and ham.

The Fair is also a place of entertainment. At the first fair, visitors could gawk at a four thousand pound steer or lunch at Williams and Overstreet's dinner tent. Early fairgoers could see the cannibals at the Igorrote Village, marvel at the "alligator boy" at the sideshow, or shriek in terror in the funhouse. Today, visitors can cheer the latest singers and gorge on pineapple whip, corn dogs, taffy, funnel cakes, and cotton candy.

Above, right: A proud winner collected these ribbons at an early fair. (V. B.)

Miscellaneous souvenirs from past fairs.

A family enjoys lunch
on the fairgrounds.
(M. S. A.)

When some predicted the demise of the state fair during the agriculture
crisis of the 1980s, Perrin noted, "State fairs
will be around, and they will embody most
of the best and a little of the worst in
American history." The Missouri State
Fair celebrated its one-hundredth fair in
2002, and promises to continue for at least
another century.

CHAPTER 1

ENTHUSIASM AND OPPOSITION:

The Beginning

Sedalia Water Company was able to provide water to the fairgrounds. (S.P.L.)

O N JANUARY 5, 1899, GOVERNOR LON STEPHENS ADDRESSED the Missouri General Assembly, encouraging the establishment of a state fair. He noted that Missouri's neighboring states were "not so favorably suited as Missouri for the production of high class agricultural products," nor did they have the "famous prize-winning herds of livestock," nor were they as financially stable as Missouri. However, other states spent thousands of dollars to maintain their state fairs and thus encourage the improvement of agriculture in their states.

Stephens acknowledged the understanding of fairs common at the time—that fairs, through the competition for prizes awarded to high quality agricultural products, would teach farmers and stock breeders better techniques and awaken in them a desire to emulate prize-winning agriculturists. The benefits of improved agriculture, Stephens said, would extend to all citizens: "A state fair will go far to awaken a slumbering interest and to benefit the industrial

classes along all lines, in stimulat-
ing a healthful rivalry in the raising of fine stock and poultry,
better and more corn, wheat, rye,
oats, barley, cotton, potatoes, etc.
to the acre; bigger and redder
apples, and in the introduction of
improved, up-to-date farming
methods, to say nothing of its
social and business features."

Stephens was not alone in his
desire for a state fair. The
Missouri Swine Breeders
Association had passed a resolution advocating a state fair in 1897, as had
the Missouri Horse Breeders Association and the Missouri State Poultry
Association. Members of these groups and other Missouri stockbreeders had
exhibited their winning animals at the 1893 Columbian Exposition in Chicago
and at the Trans-Mississippi Exposition in Omaha in 1898. The stockmen's
publication *Drovers' Journal* recognized that Missourians had won seventy-
five percent of the prizes on livestock shown at major exhibitions for the

last twenty-five years. While
farmers and stockmen showed
products at Missouri's forty-one
county fairs and at state fairs in
other states, many felt the need
for a state fair in a centrally
located rural area of Missouri.

Missouri had attempted a
state fair in previous years, but
these fairs had either closed or failed to meet the needs of Missouri's farmers.
In 1853, a state fair had been established in Boonville, but this exhibition
only lasted two years. In the 1870s and 1880s, a fair in Sedalia had billed
itself as a "state fair" under the auspices of the Missouri State Fair
Association, though it was only a local exhibition. The St. Louis Exposition,
established in 1856, was, by the 1890s, plagued by financial difficulties and
quarrelling among its board members. The *Drover's Journal* described it as
"purely local" in its make-up, in part because it did not attract small farmers,
many of whom "dislike the noise and excitement of a large city." Others
believed the St. Louis fair focused more on industry than on agriculture; the
Breeders' Gazette noted, "how little it serves as an exposition of the agriculture and livestock possibilities of the state."

On January 23, 1899, Cyrus Clark introduced House Bill 279 "to provide
for the establishment of a state fair and to regulate the control and management thereof." Clark's bill encountered "bitter and even virulent" opposition,

Governor Lon Stephens supported the establishment of
the Missouri State Fair.
(M. S. A.)

Sedalia held a fair at
Association Park in the
1870s and 1880s. (C. W.)

according to the *Drover's Journal*. Some members of the House Committee on Agriculture felt the benefits of a state fair would not justify the cost to the state's taxpayers. Other opponents feared that some parts of the state would benefit more than others from a state fair, though all would be taxed to support a fair. Those opposed to gambling disapproved of the betting on the horse races that were an integral part of fairs. Some questioned why Governor Stephens would support state spending for a fair, as the frugal Stephens was reportedly reluctant to authorize spending of state monies. As a result of questions raised about financial support for the fair, House Bill 279 was amended to include a provision forbidding the legislature from appropriating money from the general revenue account for the support, maintenance, improvement, or premiums for the fair. The bill passed the house on April 5, 1899, with a vote of eighty-eight for and twenty-one against. The bill passed the Senate on April 14, and Governor Stephens signed the bill on April 19, 1899.

The question of how to finance a state fair continued to create controversy in the legislature. Clark introduced House Bill 880, which provided that money in the Horse Breeders' Fund be used for the maintenance of the state fair. The Breeder's Fund, established in 1897, provided a way to license and regulate racetracks. Horse racing was intended, so proponents of the sport said, to encourage the breeding and development of a superior animal. Racing was the best way to test the animal's speed and endurance; that betting took place on the races simply encouraged interest in fine horses. The Breeder's Fund bill called for pool sellers and bookmakers to be licensed and to pay a fee of two dollars per day each day races were held. The money raised would be used "for the improvement of the several breeds of horses" and after a few years enough money would accumulate to help improve all livestock and agricultural products. In addition to raising money for Missouri agriculture, the bill would eliminate illegal races, such as winter races and night races, which did not encourage the display of fine horses but only provided an avenue for gambling. In addition, the bill would control the criminal element that congregated at the unlicensed racetracks. In effect, the bill legalized betting on horse races.

House Bill 880 met opposition from some of those who had voted against the establishment of a fair and from representatives of St. Louis who wished the Horse Breeders' Fund money to be used to help the racing industry in their area. An even stranger sort of opposition arose after several senators signed a petition asking that Frank James, brother of outlaw Jesse James, be appointed to serve as starter for the races. The *Moberly Evening Democrat* printed the report of a Chillicothe minister denouncing the fair as giving "license to gambling and train robbing." According to the *Chillicothe Mail and Star*, the Republican members of the House protested James' possible appointment so vigorously that Clark and other allies of the state fair had to "hustle" to ensure adequate votes to pass the bill.

Once the fair had been authorized, it was designated as a part of the State Board of Agriculture. Eighteen men were chosen to serve as the State Fair Board of Directors. D. A. Ely of Sublette was elected president and J. R. Rippey of Glenwood was secretary. Other members of the committee included J. W. Hill of Chillicothe; Alex Maitland of Richmond; Eugene Rhoades of Fairfax; J. F. Groves of Corder; George B. Ellis of Appleton City; Nicholas Gentry of Sedalia; L. F. Luthy of Lebanon; J. A. Potts of Mexico; Thomas B. North of Gray Summit; Norman J. Colman of St. Louis; W. R. Wilkinson of St. Louis; Charles M. O'Connell of Fredericktown; W. J. Roberts of Oak Ridge; and C. P. Cook of Round Grove. Governor Lon Stephens, Super-intendent of Schools W. T. Carrington of Jefferson City, and Dr. H. J. Waters, dean of the University of Missouri School of Agriculture in Columbia served as ex-officio members.

The Directors set about choosing a location. House Bill 279 stipulated that the fair be located in a rural area in the central part of the state, in a town easily accessible to exhibitors and visitors. In the late nineteenth and early twentieth centuries, when the state proposed a new state facility, communities wishing to be chosen as the location of that facility were required to

In 1896, the M.S.& T. Railroad built this depot in Sedalia. (S. A. C. C.)

give the state land on which to build the facility and to guarantee services such as water, electricity, and perhaps an infrastructure of roads and sidewalks. The state fair directors asked that the communities be able to give the state a plot of land of at least one hundred acres suitable for the development of a permanent site for a state fair, and to describe in their presentation the amenities being offered.

Geographer Fred Kniffen, who has studied the location and ground plans of fairs throughout the United States, points out that long distance transportation facilities are an important determinant of the location of a fair, as exhibitors and visitors must be able to travel to the fair easily. He further notes that since fairs are usually on the outskirts of town, local transportation is also essential. In addition, a city's ability to feed and house visitors becomes a necessary part of the location. The Directors were to examine the proposals presented by each town that wished to be considered as the site of the state fair, to visit each town, and to vote to select their choice. Six communities—Centralia, Marshall, Mexico, Chillicothe, Moberly, and Sedalia—competed for the prize they believed would encourage civic growth and enhance their economy.

The towns competing showed their suitability as the site of the state fair by demonstrating their civic pride with banners and brass bands. They boasted of the modern amenities their cities had to offer. Each brought its best orators to speak in its favor. Local newspapers engaged in a war of words, praising their own towns and belittling the others. Several towns displayed the importance of agriculture to the region by showing off farms known for producing purebred stock. Others, proud of their extensive manufacturing plants, offered tours of them. Each town tried to outdo the other, not only in what it offered the state for the fairgrounds, but also in the friendliness, food, and drink it extended to the Directors.

Marshall, a city of approximately five thousand, served as the county seat of Saline County. Home to Missouri Valley College, it boasted sixteen churches, schools for white and black children, and public water and electric service. Its four hotels and seven restaurants would, its citizens believed, be able to serve the crowds drawn by the fair. Marshall offered 160 acres plus $20,000 for the building of the grounds. It promised streetcar service to the grounds, electricity on the grounds, and advertised the paved roads that would make the fair more easily accessible. Marshall's supporters entertained the Directors lavishly, serving them a dinner costing three dollars per plate, an amount equaling two days' wages for many factory and office workers. It gave

Business buildings surrounded the Saline County Courthouse in the Marshall town square at the turn of the century. (S. H. S. M.)

the Directors souvenir "badges as big as handkerchiefs," and sent them on to the next site on a train car loaded with wine, beer, and cigars.

Moberly, called the Magic City, was a railroad town located on the Wabash, St. Louis, and Pacific Railroad, with connections to the Iowa and Minnesota railroad systems. The city of eight thousand was home to the Wabash Railroad shops, and had public water, telephone, and gas and electric lighting. Moberly, with the large transient population typical of a rail center, had four large hotels and several smaller hotels. Moberly brought representatives from the surrounding towns of Salisbury, Macon, Palmyra, and New Franklin to speak on its behalf.

The Moberly citizens treated the Directors royally, leading tours of the Wabash Railroad shops and the Blees stables, home of fine racing horses. Following a lavish dinner at the Merchants' Hotel, they toured the Miller's Fair Ground Park and heard several speeches. They then visited the Sandison Brick Plant, the Jacoby Foundry, Faesler's Machine Shop, the Wabash Railroad Hospital, and Goetze's Conservatory of Music. Their tour was made easier by the "miles of paved streets" of which the city was justly proud. While on the tour they were serenaded by the Bachelor's Band. The Directors ate supper at the homes of local citizens, and then returned to the hotel for a "smoker" and a concert by the Goetze Orchestra.

Mexico, the county seat of Audrain County, was also a progressive city with a population of nearly five thousand. It boasted a hospital, a fire department, and public schools for black and white children. Mexico had four restaurants, seven hotels, and two saloons, enabling it to serve visitors adequately. Mexico offered the state a 124-acre improved fairgrounds, or 160 unimproved acres with the use of the present fairgrounds for three years. The

Reed Street in Moberly boasted many businesses. (S. H. S. M.)

Chicago and Alton Railroad promised to extend its lines to the fairgrounds. The Directors visited a fairgrounds site in nearby Callaway County, ate a huge dinner complete with wines and mint juleps, heard speeches in Mexico's favor, and then toured the other sites and the city.

Chillicothe, county seat of Livingston County, was already the site of the State Industrial School for Girls. The town of approximately 6,500 was a railroad town, located on the Chicago, Milwaukee, and St. Paul Railroad, the Hannibal and St. Joseph Railroad, and the Wabash Railroad. City services included electricity provided by two electric companies, hospitality offered by six restaurants, nine hotels, thirteen saloons, and transportation on a street railway. Chillicothe promised land plus $15,000 and treated the visitors to a tour of the Weavergrace Hereford farm. Chillicothe put on its "best appearance" for the committee, and feasted them liberally at the New York Store clubrooms. Following a demonstration run by its fire department, a band concert in Elm Park entertained the Directors.

Centralia, with a population of 1,700 the smallest of the communities, was located at the intersection of the Chicago and Alton Railroad and the Wabash and Columbia Branch Railroad. Although the town had electric service, it lacked other amenities. Centralia supporters believed its location made the city the most suitable; it was closest to the University of Missouri College of Agriculture in Columbia. Centralia promised a 120-acre fairgrounds and $10,000. Centralia, decorated with bunting, entertained and fed the directors

Mexico's Jefferson Street was shaded with trees.
(S. H. S. M.)

Chillicothe's town square highlighted its business district. (S. H. S. M.)

at a banquet at the Ringo Hotel, but supported temperance, and so was the only town that did not include wine or beer as part of the refreshments served. Centralia would later make an issue of its temperance stance.

Sedalia offered 160 acres of land set aside in 1896 as the site for the proposed state capital with use of the existing fairgrounds at Liberty Park if needed. The Missouri Pacific and M.K.& T. Railroads intersected in Sedalia, and a branch line to Lexington and the Sedalia, Warsaw and Southern Railroad met the Missouri Pacific there. The city water company promised water service to the fairgrounds, the electric company promised electric lighting, and the street car company promised trolley lines and a percentage of the income from the line to the fairgrounds, an amount predicted at $15,000 to $20,000 over a ten-year period.

Sedalia had not published a great deal of information about its proposal prior to the visit, and other competing towns did not believe Sedalia to be interested. Sedalians had, however, worked diligently to secure the amenities needed. Its proposal to the Directors included letters from the electric light company, the water company, the Missouri Pacific Railroad, and the street car company verifying their services. Other towns did not provide as much documentation of their offerings, suggesting that Sedalia's thoroughness may have been a factor in the Directors' decision.

The very tired and exceedingly well-fed Directors returned to Jefferson City on June 3, 1899, accompanied by a band from Mexico which played at each stop and played "Dixie" when Governor Stephens appeared. The committee heard more speeches by representatives of each town. They then

A bank dominated Allen Street in Centralia. (S. H. S. M.)

Below, right: Carriages and streetcars provided transportation on Sedalia's Ohio Street. (P. C. H. S.)

began a voting process that would last into the night. After ten ballots, one town secured a majority. The state fair would be in Sedalia.

Sedalia's representatives were not on hand to hear the results; they had left Jefferson City on the 11:55 train. News reached them by telegraph at California, Missouri, and the train made an unexpected stop. The Sedalia contingent was, said the *Sedalia Democrat*, a "delighted, disorderly, almost delirious delegation."

Almost immediately, the other towns began to question the Directors' choice in their newspapers. Chillicothe recognized that its location in the north of the state had worked against it. Centralia, which had received the fewest votes, only one vote in each balloting, called the selection process a "confidence game," claiming the Directors had decided on Sedalia before the tours, and intimating the Directors had "not shown the dignity and deliberation expected of a representative body," having been unduly influenced by the "alluring blandishments of the tempting liquid refreshments." The *Mexico Intelligencer* suggested that Sedalia won because of luck, but implied that something was amiss: "While we are somewhat curious as to how you turned the trick, we will ask you no embarrassing questions." The *Moberly Evening Democrat* suggested political finagling had resulted in Moberly being "buncoed out," and complained because Sedalia, a stronghold of the Democratic Party, had used a powerful Republican, John Bothwell, as its spokesperson.

The feeling that the choice of Sedalia had been a foregone conclusion may have had some merit. As early as 1872, a group of men sponsoring a local fair in Sedalia had called itself the Missouri State Fair Board, a designation affirmed by the State Board of Agriculture, which called the Sedalia Fair the Missouri State Fair. In addition, John Bothwell, who had strong ties to the Missouri Pacific Railroad, was thought to have used his influence with the railroad and its lobbyists in the legislature to influence the vote toward Sedalia. Questions about the legitimacy of Sedalia's selection would resurface when the abstract for the land Sedalia offered was found faulty. While Sedalia's representatives were fixing the problems with the abstract, the other towns continued to protest. However, the Directors remained firm in their choice, and Sedalia made plans to host the state fair.

The Missouri State Fair Board of Directors elected an executive committee to manage the operation of the fair. Norman J. Colman was president; Nicholas Gentry was vice president; J. R. Rippey was secretary; Charles McAninch of Sedalia was treasurer; and Alex Maitland and George Ellis were members. Secretary Rippey was responsible for the day-to-day management of the fair.

The question of financing the fair arose again when the actual planning and building began. The Horse Breeders' Fund was to have generated enough money to build permanent buildings on the fairgrounds. Secretary Rippey anticipated the fund would contain between $20,000 and $25,000 by 1900. Unfortunately, the estimates of its receipts were grossly exaggerated. On January 8, 1900, the State Federation of Labor requested the General Assembly to appropriate money for the fairgrounds so as to "encourage the development . . . of the Agricultural, Horticultural, Mechanical, Mineral, Stock Raising, and all other industrial interests of the state." By September 1900, the Horse Breeders' Fund contained only $14,277. The State Industrial Association, a group of agriculturalists and industrialists, resolved in December 1900 that the state should "make ample appropriation for the creditable equipment" of its state fairgrounds. They were joined by members of the Missouri State Poultry Association, which asked for $25,000 to construct a poultry building on the grounds.

On January 11, 1901, just three days before he left office, Governor Stephens addressed the General Assembly, stressing that a liberal appropriation was necessary if the fair were to have the permanent buildings that Missouri's agricultural and industrial resources deserved. Secretary Rippey and Director Alex Maitland asked the House Appropriations Committee to consider providing the fair with adequate funds. Stockman Charles Leonard of Bell Air organized three groups of stockmen to lobby the legislature on behalf of the appropriation. A committee of house and senate members

John Bothwell championed the state fair before the General Assemly. (B. L. H. S.)

convened to consider the appropriation suggested that $100,000 be set aside for the fair. Pettis County Representative John Bothwell, Mr. Martin, Former Senator Charles Yeater, Secretary Rippey, the Honorable O. M. Barnett, and Colonel T. F. Mitchum supported the committee's request for $100,000, but the Senate reduced the fair appropriation to $50,000.

After a series of committee meetings, the house debated the issue. In a rather spirited discussion, Representative W. D. Dalzell of Webster County noted that the appropriation was contrary to the original law authorizing a state fair. J. Marion Welker of Bollinger County, J. N. Gipson of Chariton County, and J. W. Farley of Platte County also opposed the appropriation. Dalzell objected further, raising the issue of gambling on the horse races as another opposition to the fair, declaring the fair to be "the only legally authorized gambling resort in the state." Bothwell pleaded for money for the fair, noting that Missouri had already appropriated over one million dollars for exhibits at the Pan American Exposition and for the St. Louis World's Fair. The appropriation bill finally passed on March 18, 1901, but the appropriation of $50,000 seemed to many stockbreeder and farmer's journals to be "limited" and "insufficient" and "insignificant."

Governor Dockery was expected to sign the bill, but delayed because the state treasury was allegedly nearly empty. The Missouri State Brewers' Association had protested the state-mandated inspection fee on barrels of beer, and had taken its case to the courts. While the state supreme court was reviewing the issue of inspection fees, the Brewers' Association had refused to pay what the state deemed it owed. On April 15, 1901, the court ruled, and the brewers paid $191,250 into the state's depleted treasury. Governor Dockery signed the appropriations bill.

Contractor James A. Heck graded the racetrack. (C. T. and A. J. H.)

The fair now had the money needed to continue work started the previous year. During 1900, Directors Colman, Gentry, Maitland, Ellis, and Rippey had visited state fairgrounds in Ohio, Illinois, and Indiana to learn appropriate layouts for a state fair. The directors ordered preliminary work done on the grounds using money from the Horse Breeders' Fund. Surveyor T. O. Stanley viewed the site, identified possible locations for the racetrack, and made estimates of construction costs. Kansas City landscape architect George Kessler, probably with the help of his employee Henry Wright, drew a bird's eye view of the grounds. Sedalia architect Thomas W. Bast designed twenty-one buildings, which he estimated would cost $284,450.

Norman Colman

When Missouri decided to hold a state fair, it turned to Norman Colman as the obvious man to head the planning process. Colman, publisher of the popular farm periodical *Colman's Rural World* and active in politics, was widely known by Missouri farmers.

Norman Colman, born in 1827, grew up on a farm near Richfield Springs, New York. He studied law at the University of Louisville, graduating in 1852. After practicing law in Indiana for three years, he moved to St. Louis. He bought a farm near the city, began experimenting with scientific methods of farming, and became involved in local politics, serving as a member of the St. Louis Board of Aldermen. He remained active in politics, campaigning on behalf of numerous Democratic candidates. In 1865, Colman won a position on the Missouri State Board of Agriculture, and from 1866-67 he served a term in the General Assembly.

After the Civil War, Colman created *Colman's Rural World*, an agricultural periodical that remained popular and influential throughout the fifty years it was published. In 1874, he began a term as lieutenant governor. President Grover Cleveland named Colman the United States Commissioner of Agriculture in 1885. Colman worked hard for the passage of the Hatch Act in 1887, which authorized the establishment of agricultural experiment stations by the land grant universities.

Prominent Men and Women of the Day describes Colman as a "thorough and practical agriculturalist." His duties as Commissioner of Agriculture included disseminating information about the most effective agricultural methods, and collecting, testing, and distributing seeds. In 1887, he advocated a "highly creditable" method of eliminating pleuropneumonia.

When the Department of Agriculture became a cabinet position in 1889, President Grover Cleveland appointed Colman Secretary of Agriculture. Historian Pat Curran notes that Colman served at the time when the United States changed from being primarily agricultural to being a predominantly urban nation, with the "increasing demand for fewer farmers to produce ever greater quantities of food."

Following his term in Washington, Colman returned to Missouri, where he continued to serve on the State Board of Agriculture and as a member of the University of Missouri Board of Curators. He was appointed to the Missouri State Fair Board in 1899 and was influential in the development of the Missouri State Fair as a forum for agricultural education.

The stables for race horses were the first buildings erected. (V. B.)

Work started in late summer of 1900. Contractors George Menefee and James A. Heck of Sedalia had contracted to construct and grade the racetrack, a task made more difficult by the location chosen by the directors. A Western Wagon Loader and Elevating Grader pulled by sixteen horses kept ten teamsters busy hauling the dirt to build up the north side of the track and level the south side. At the same time, the Missouri Pacific Railroad had engaged a force to excavate for the spur line to be built onto the fairgrounds. On September 11, 1900, the *Sedalia Democrat* reported that the directors authorized the building of a stable for racehorses with sleeping quarters in the attic for attendants at the eastern entrance to the fairgrounds, and a barn for cattle and another for exhibition horses on the southwestern part of the grounds south of the grandstand. In January 1901, Governor Stephens announced that H. M. Hammond had completed these three buildings.

After the appropriations bill finally passed in March 1901, work began again on the grounds. Secretary Rippey faced the daunting task of changing a "bleak prairie" into the "neatest and most picturesque fairgrounds in the West." The racetrack needed an additional $8,000 worth of work, which was to be completed by August 1. On March 27, 1901, the directors authorized Thomas H. Johnson and Butler to build ten frame barns to house racehorses at a total cost of $5,540 and a five-hundred pen swine and sheep barn for $7,994. Joseph T. Heckert and S. Wilson Ricketts were to build a second 120-foot by 60-foot frame cattle barn to hold ninety head for $2,755, and A. J. Hutchinson a second exhibition horse barn with spacious stalls for $3,075. Heckert and Ricketts also won the contract for construction of a two-story, forty-by-forty-foot administration building for $1,876 and a four-room custodian's cottage for $687. The Midland Bridge Company of Kansas City was to build an eighty-foot, $7,085 section of the planned $30,000 grandstand. Kessler laid out drives and walkways.

The *Kansas Farmer* noted that fair officials "labored under difficulties that seemed almost impossible to surmount." The pressure of time—only five months to seek bids, award contracts, and finish buildings—was complicated by severe weather. According to A. E. Hackett of the U.S. Weather Bureau, the spring was "cold, stormy and disagreeable" with snow and heavy rain. May, June, July, and August were extremely dry, with precipitation about one-fourth the normal amount. In addition, the summer was the longest period of extreme heat to that date recorded in Missouri; temperatures rose over one hundred degrees during July and August.

Throughout August, Sedalia newspapers reported on construction at the fairgrounds, optimistic but perhaps not very realistic about the ability of contractors to have everything completed by the fair's opening on September 9.

A trolley line extended from downtown Sedalia to the fair-grounds in 1903. (V. B.)

Below: Heck and Menefee pur-chased a Contractor's Dumpwagon for use on the racetrack. (C. T. and A. J. H.)

Menefee and Heck worked very slowly at grading the racetrack because of the heavy rain during the spring and because of the hard, dry soil and dust of the summer. They fell so far behind in construction of the track that Kessler ordered them to secure another grading machine and to work ten-hour days in order to complete the track. In addition, they were forced to drill a well to provide water for their animals, their workers, and for sprinkling the track.

Contractors were further hampered by a strike of the Amalgamated Steel Unions. The Sedalia Water Company had promised water service to the grounds, but the pipes could not be shipped because of the strike. When the materials finally arrived in mid-August, the water department completed the piping, and a hose was arranged to enable the contractors to water the track. By opening day, the Sedalia Water Company had provided water service and fire hydrants to the grounds.

The steel strike also affected the construction of the grandstand, as Midland Bridge Company could not receive the necessary materials. On August 8, 1901, the date their contract was to expire, materials for the grandstand had not arrived, so Midland forfeited its contract, but was granted the privilege of applying for the contract the next year. When the steel ordered by Midland finally arrived at the fairgrounds, fair custodian W. S. Allison would not allow Midland to unload and store the steel on the grounds. Eventually the dispute was resolved, and the *Kansas Farmer* reported that "tons upon tons of the steel structural material" lay on the grounds in preparation for con-

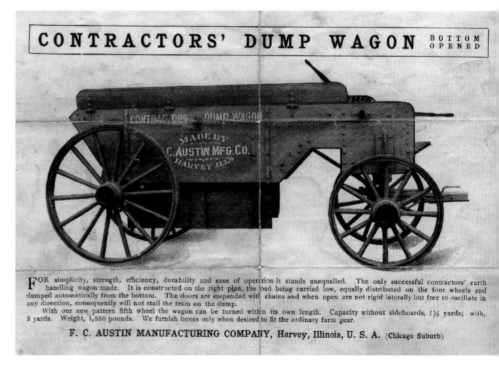

CONTRACTORS' DUMP WAGON BOTTOM OPENED

For simplicity, strength, efficiency, durability and ease of operation it stands unequalled. The only successful contractors' earth handling wagon made. It is constructed on the right plan, the load being carried low, equally distributed on the four wheels and dumped automatically from the bottom. The doors are suspended with chains and when open are not rigid laterally but free to oscillate in any direction, consequently will not stall the team on the dump.
With our new pattern fifth wheel the wagon can be turned within its own length. Capacity without sideboards, 1½ yards; with, 2 yards. Weight, 1,550 pounds. We furnish boxes only when desired to fit the ordinary farm gear.

F. C. AUSTIN MANUFACTURING COMPANY, Harvey, Illinois, U. S. A. (Chicago Suburb)

A   H I S T O R Y   O F   T H E   M I S S O U R I   S T A T E   F A I R   **29**

Trolleys brought visitors from downtown Selalia to the fairgrounds in 1903. (T. P.)

struction. H. M. Hammond rushed to build a temporary wooden grandstand.

Financial difficulties prevented the Sedalia Electric Railway Company from extending its lines onto the grounds. The company, which had overspent in the previous year, had declared bankruptcy and been sold. The M.K.& T. and the Missouri Pacific Railroads agreed to run extra trains to the fairgrounds from their downtown depots; both railroads constructed loading platforms near the livestock buildings and depots for passengers on the grounds.

Other aspects of preparation seemed to be under control. Architect Bast served as superintendent of the grounds, responsible for the location of exhibits and concessions. Fair Directors elected Colonel Ira Hinsdale, a "broad-minded gentleman" who would permit "no guilty man to escape," to be chief of the fairgrounds police. The *Sedalia Democrat* praised Hinsdale's selection, assumed the crowds would bring crime, and encouraged the police and courts to do everything necessary to maintain order. Sedalia Fire Chief

Willis arranged to move an engine and a force of firefighters to the grounds. Dr. William Shirk, a Sedalia physician and assistant surgeon of the Second Regiment Home Guard, planned to set up and staff an emergency hospital near the Administration Building. Ladies of the Woman's Christian Temperance Union had agreed to assist Shirk, and had erected three tents in which to provide a day nursery where small children could be left while their parents toured the fair and a resting place for tired women. The Knights and Ladies of Service would serve ice water to visitors, and the WCTU would serve chocolate and tea. The *Democrat* reprimanded the Pettis County Court for not offering its courthouse lawn benches to augment the chairs provided by the Knight and Ladies of Security and the WCTU.

By September 1, the fair facilities seemed ready. Telephone workers completed the telephone connections. A Western Union office, staffed by Mae Hunicka, offered telegraph service, and Walter Kennedy managed the State

Fairgrounds Post Office branch. James Leake established an Information Booth on the grounds. The Sedalia Awning and Mattress Company erected two tents to use for implement displays. One of the cattle barns had been set aside for use as an exhibition hall, as had one of the horse barns, which was used for textile and fabric exhibits. A large tent had been erected to house dairy cattle. The sheep and swine pavilion, praised by the *Breeder's Gazette* as "the most imposing structure on the grounds," was "nearly 200 feet square, with 500 pens, and in the center of it is the show arena 50 foot square, with raised seats for 1200." The mile racetrack had been repeatedly sprinkled and graded, and was ready for the horse, automobile, and bicycle races. The *Kansas Farmer* noted that the track was the best in Missouri and one of the chief attractions of the fair. The large barns for stabling racehorses to the east of the track were completed. Workers finished erecting tents to house the poultry exhibits.

Charles McAninch, manager of the ticket department, named T. S. Hopkins to be his assistant, and appointed Ollie Terry of LaMonte superintendent of tickets. Sedalians W. O. B. Dixon, Sam A. Rosse, George M. Weed, Dr. T. P. McCluney, Douglass Lane, Fred Hoffman, and Hughesville resident Henry Schmidt were ticket sellers. Gatekeepers were O. A. Stine of Dresden, T. G. Kelly of Hughesville, Lawrence Bothwell and John Baker of Sedalia, E. E. Williams of Georgetown, Walker Redd of Dresden, and James Elliott of south of Sedalia. Two men were appointed to purchase tickets from the sellers and deliver them to those patrons who did not want to get out of their carriages.

While the fairgrounds may have been ready, Sedalia was not. On September 1, prompted by Fair Executive Committee Vice-President N. H. Gentry's complaints that Sedalians did not seem concerned about hosting the anticipated crowds, Mayor John Babcock called a citizens meeting and appointed six committees to organize hospitality for fair visitors. W. M. Johns chaired the committee responsible for information and rooms. Member Melvin Haley set up headquarters downtown and on September 6, just three days before the fair was to open, sent canvassers door to door to solicit rooms in private homes that might be rented during the fair. Ultimately, his canvassers identified three hundred

Sedalia firefighters maintained the state fair fire department. (S. P. L.)

The first Sheep and Swine Pavilion was built in 1901. (M. S. F.)

available rooms. Haley prepared a list of rooms and established a Bureau of Information in downtown Sedalia that could direct visitors to hotels, boarding houses, and homes, and hired boys to work as guides. Calvary Episcopal Church ladies served food in the Bank of Commerce Building at 125 South Ohio, and Sacred Heart ladies served food at the O'Meara Building at 320 South Ohio.

Fair Directors solicited local vendors to provide food on the grounds. Originally they had intended to ask $3,000 for exclusive rights to food concessions; when no one had accepted the offer by August 21, they dropped the price to $1,800. By early September, so few concessionaires had applied that the Directors eliminated the offer of an exclusive contract. Seymore Williams and D. M. Overstreet had erected a tent seating 104, plus a separate seating area for African-American customers, for their dining hall, but fair Directors feared they would not be able to handle the crowds. The Directors asked churches to serve food, but few seemed interested. The *Sedalia Democrat* reported that plenty of people would be selling pies and cakes, but begged an enterprising person to sell more substantial food—sandwiches and coffee. Fair officials were especially worried that concessionaires would not be able to handle the anticipated crowd of fifteen thousand on Wednesday, World's Fair Day.

The week before the fair saw frantic activity on the grounds. Building superintendents housed livestock and arranged displays. On September 8, the rains came. Winds nearing 130 miles per hour tore through southwest Sedalia, damaging the tent set aside for dairy cattle and almost injuring a man and his team then in the barn. The rain settled the dust and left puddles on the walkways. On Monday, September 9, the first Missouri State Fair opened.

Sedalia at the Turn of the

Sedalia, the county seat of Pettis County, advertised itself as a "live" city, meaning that it was progressive and modern. Sedalia was also a lively city. Ragtime music, given an impetus by the publication of Scott Joplin's "Maple Leaf Rag" in Sedalia in 1900, wafted from the open doors of the city's twenty-nine saloons and from the open windows of proper middle-class homes. Trains roared through, bringing goods, commercial travelers, and hoboes, and taking livestock, grain, and people to exotic locations. Sedalia's reputation as "wide open" meant that gambling halls and its saloons thrived. Main Street was called "Battle Row" because of the fights there. Sedalia also, unfortunately, had a reputation for government corruption, as its mayors pretended Sedalia's wilder side did not exist.

Divisions between the respectable and not-so-respectable marked much of the city's political activity. Sedalia was divided in other ways as well—by race, by ethnicity, and by social and economic class. Sedalia's African-American community supported George R. Smith College and Lincoln School, many businesses, and seven churches with buildings and furnishings worth over $300,000 in spite of the difficulties African-Americans faced during this time. The strong divisions based on social and economic class limited opportunities for the working class and their children. The economic divisions contributed to the development of a strong Populist Party and made Sedalia a center of labor union activity.

Sedalia saw itself as the "commercial, industrial and educational metropolis of Central Missouri.... a desirable place of residence as well as a manufacturing and distributing center." Its location at the intersection of the Missouri Pacific and the Missouri, Kansas & Texas Railroads made it a center of business, industry, and the wholesale trade. In 1898, gross profit from these firms totaled over 5.5 million dollars. Sedalia's population, according to the 1900 census, was 15,231. The city boasted a complete range of services—a water supply, a sewer system, electric service, a streetcar with twelve miles of track, streetlights, a gas plant, thirteen miles of paved streets, a twelve-man police force, and a well-equipped fire department.

The M.K. & T. Railroad in 1898 opened a new shop for the manufacture and repair of railroad cars. The Katy's general offices were in Sedalia, as was the company's hospital. In 1903, the city's employment base would be further enhanced by the location of the Missouri Pacific shops that would employ 1,800 people. Sedalia continued to solicit new businesses and industry by publishing the advantages Sedalia had to offer.

Sedalia provided public schools for both black and white children, and was home to the George R. Smith College for Negroes, the Central Business College, and Hill's Business College. Sedalia's seven newspapers— the *Democrat*, the *Sentinel*, the *Capital*, the *Bazoo*, the *Times*, the *Journal*, and *Rosa Pearle's Paper*—reflected the views of Democrats, Republicans, African-Americans, German-Americans, and society ladies. The Carnegie Library, the first in Missouri, opened in 1901. Sedalians maintained their spiritual connections by attending the city's two Catholic churches, eighteen Protestant churches, a Jewish synagogue, or the active Salvation Army.

Sedalia's prosperity may, however, have been an illusion. Sedalia had suffered major economic losses during the 1890s as a result of the national depression. A more serious blow to the local economy came in 1894 when the First National Bank, already suffering from a series of unwise investments, collapsed after its cashier stole the contents of the bank's vault and fled to Mexico. In 1899, the city mounted an intensive campaign to collect past due business licenses and fines so that it could pay its police force. Sedalians believed that achieving the Missouri State Fair would help the city's economy by creating jobs and by bringing visitors to Sedalia.

Sedalia had a history of struggling to be the site of a state facility and then not being awarded the prize. In the 1870s, Sedalia had come close to being chosen as the location of the state normal school, now Central Missouri State University, but had lost in a change of legislation many thought suspicious. Sedalia newspapers suggested in the late 1870s and early 1880s that Sedalia push to become the site of a new "lunatic and inebriate asylum" or a school for the blind. In 1892, after the fire that destroyed the main building on the campus of the University of Missouri, Sedalia encouraged the University to relocate. During the 1880s and 1890s, Sedalia campaigned to have the state capital moved from Jefferson City, citing better railroad ser-vice and better roads; the measure was put to a state-wide vote in 1896 and failed miserably. Certainly Sedalia felt it deserved a state facility.

Sedalia was, in most respects, a typical small city of its time, carefully hiding its negative elements, basking in the warmth of civic pride, and optimistically hoping to grow and succeed.

CHAPTER 2

"RADIANT IN HOLIDAY ATTIRE":

The First Fair

"THE USUAL AND UNAVOIDABLE CONFUSION OF THE FIRST DAY OF
a great exposition prevailed," wrote a *Sedalia Democrat* reporter.
Secretary Rippey and his assistants answered questions, placated quarrelsome patrons, and "handled the work that crowded upon them in an admirable
and expeditious manner." The *Missouri State Fair Premium List* specified that
Grounds Superintendent Thomas Bast was to "see that visitors do not violate
regulations" and that "no space is occupied by a stand of any character except
such as have purchased and paid for the privilege . . . and are complying
strictly with the terms of the concession granted." Bast, the "busiest man on
the grounds," spent the morning "continuously surrounded by applicants for
privileges and concessions."

Department superintendents were equally busy. Their task, according to the

Early fairs were crowded
with visitors. (T. P.)

Premium List, was "to direct the arrangement of all articles or animals on exhibition." Each was also responsible for "caring for and returning the entry book of his department" with winners identified and for submitting a final report at fair's end. Fair regulations allowed exhibitors to bring items for competition or display before noon on Monday, and some departments were still grappling with the details of receiving and placing entries. By noon, such details were almost finished; the *Sedalia Evening Sentinel* acknowledged that after the shows and stands were set up, "the grounds

"A Prize Winning Exhibit at Sedalia"

County exhibits allowed individual counties to boast about their agricultural output. (M. S. F.)

have a very business-like appearance." The *Sedalia Democrat* promised on Tuesday a "Missouri State Fair . . . radiant in holiday attire."

Difficulties in arranging exhibits occurred in part because the number of exhibits was far more than anticipated. Poultry Superintendent Codding, realizing that chickens are not usually in show condition in September, had not expected the large numbers of birds that materialized. Horticulturists and orchardists brought far more carefully nurtured fruits and vegetables than had been expected, considering the drought. Textile Superintendent J. M. Cannon sent his force of assistants to find additional space for the hundreds of items of needlework that arrived Monday. The fair promised to be "larger and better" than anticipated, its "magnitude," bragged the *Sedalia Democrat*, "seen in the excellent and numerous exhibits in each department."

The fifteen-piece New Bloomfield Band, directed by C. J. Howertine, arrived in Sedalia from Callaway County on the Katy train. Ignoring the custom of traveling musicians to play a few numbers at the depot, they quickly boarded a train for the fairgrounds. Resplendent in their elaborate red and gold uniforms, they marched to the fair's bandstand and played the rousing marches and popular hits of the day. Street musicians with hand organs and harps added what a *Democrat* reporter satirically called "melodious strains" to the festive cacophony. Children, admitted free on Children's Day, came by the hundreds, and gazed in wonder at the exhibits, some of which must have seemed miraculous.

The first fair hosted an exhibition of oddities—not a true freak show, but oddities nonetheless. A 4,000-pound steer amazed onlookers. A group of "snake eaters" awed the crowds with their ability to handle poisonous reptiles. Since the fair was touted as an educational exposition, even the oddities had an educational component. For example, exhibits of indigenous people from exotic locales were popular at turn-of-the-century fairs, for they educated the public about other ways of life while reinforcing the prejudices that suggested a distance between backward, uncivilized people and the progressive, civilized United States. The Australian Colony, housed in a tent, offered fairgoers a glimpse of life in that mysterious South Pacific continent.

A more "scientific" educational exhibit displayed x-ray illustrations of broken bones and malformations from the World's Institute of X-Ray Diagnosis and Treatment of St. Louis, enabling fairgoers to learn about the human body and demonstrating the degree to which x-rays had become an "indispensable adjunct to the healing art." In overstatement perhaps more suited to a sideshow barker, the x-ray laboratory announced it could cure "skin cancer, lupus, ulcers, and all skin diseases" in the patient's home "by the use of storage batteries."

The Fair Directors had prepared a program

pamphlet advertising the fair and extolling the wonders a visitor could see. The exaggerations of the program fit the enthusiastic nature of a fair. The theme of the first fair—"profitable fun for all who come"—emphasized the benefits of agriculture education to Missouri's farmers. Those hoping to learn about new methods, crops, and equipment could visit a large display by the University of Missouri School of Agriculture. Lecturer T. I. Mairs demonstrated soja beans, a crop then not grown in Missouri, and cowpeas. Charts illustrated the comparative feed value of cowpeas to clover and timothy. Farm supply companies demonstrated their latest products. Farmers saw new varieties of fencing from the Joliet Illinois Wire Fence Company and scales from the Chicago Scale Company. The Sedalia-based Kelk Carriage Works set up a tent to display the wagons and buggies it made. The Ohio Cultivator Company showed new plows and cultivators.

Implement companies demonstrated labor-saving devices in exhibits designed to educate and to sell, as the rhetoric of their displays indicates. The

Sure Hatch Incubator Company attracted "universal attention" from the hundreds who lingered to watch chicks "pick up the thread of life with an energy and vigor very seldom found among chickens hatched in the ordinary way." The incubators, "built on scientific principles" were "made right, guaranteed for years and priced so low that anyone can purchase." The Sharples Cream Separator shown by Ed J. Chubback of Chicago, could "have the cream ready for churning and the skim milk ready for feeding" immediately after milking. The machines, the "very acme of the mechanics' art," worked in "absolute safety" and with a "minimum amount of labor."

Missouri at the turn of the century was primarily an agricultural state; over two-thirds of Missourians lived on farms. The state fair, reflecting as it did the agricultural interests of the state, created categories of exhibits to showcase what was produced on Missouri farms. The categories of exhibits tell much about what was produced not only for the commercial market, but also for home consumption.

The *Breeders Gazette* noted that Missouri had suffered from drought and its agriculture products were of "pauper quality," a comment confirmed by the August Crop Report from the State Board of Agriculture, which noted that sixty-five percent of the corn crop was "beyond recovery," only fifty percent of the oats "worth harvesting," and most fruits and garden crops "generally poor." However, the fair's exhibit was, according to the *Drover's Journal*, a "revelation to all who had supposed that Missouri had been ruined by the

A postcard extols the agricultural produce from Missouri. (D. C.)

5 - "A few Tomatoes." For the Missouri State Fair, Sedalia, Oct. 2-8.

drought. . . Equal in quality" to the produce shown at any state fair, "it evidenced both the versatility and skill of the farmer."

Pettis County exhibited its produce at an early fair. (M. S. F.)

The program pamphlet identified Missouri as the "Orchard of the Globe," a state which led "the world in variety and excellence of Agricultural Products." Almost all farm families and many town dwellers raised vegetables in garden plots; some supplemented their incomes by selling produce. Fruit growing was an important part of both commercial agriculture and home production, with apples, peaches, and grapes the most important crops.

The Horticulture and Field Crop exhibits were housed in one of the barns originally built to house cattle. The Horticulture Exhibit, under the direction of Superintendent G. A. Atwood, presented six varieties of grapes, eight varieties of pears, seven varieties of peaches, and thirty-six varieties of apples. The *Premium List's* categories of vegetables included turnips, tomatoes, parsnips, peppers, melons, beets, carrots, cucumbers, peas, radishes, rhubarb, celery, potatoes, and beans, with special classes for extra large watermelons, cabbages, and onions. Happy winners took home a total of $307 in prizes.

Corn was Missouri's most important field crop, constituting over 6.5 million acres in 1900. Wheat accounted for 2.1 million acres. Other important crops included oats, hay, tobacco, sorghum or kafir corn, cowpeas, rye, and alfalfa. Soybeans had not yet been introduced. Farmers displayed several varieties of corn, including popcorn, sweet corn, field corn, and white, mixed, and yellow dent corn, vying for total premiums of $288. The display of field crops also included red winter wheat, rye, black oats, timothy and clover seed, timothy and red clover hay, millet, orchard grass, cowpeas, lima beans, and castor beans. The display, while not large, was, said the *Kansas Farmer*, "a creditable state display for a droughty year." The *Sedalia Democrat* pronounced the exhibit "more instructive than usual. The noble fight . . . against the dry weather has brought out the latent qualities of the energetic and resourceful farmer."

The Floriculture Exhibit had not been hurt as much by the drought as had the agriculture exhibits. Under the direction of Superintendent Charles Gelvin, a Sedalia greenhouse owner, the exhibit showed off rex begonias, geraniums, cacti, ferns, and variegated foliage plants. Gelvin and Sons Nursery won eight of the awards; only one of the many awards, in the amateur's only class, was won by Mrs. J. W. Planch's begonias. Judges were prohibited by the state fair regulations from having "any personal interest . . . in any article submitted for his examination," but apparently building superintendents were exempt from this rule.

The program book announced the "greatest live stock exhibit on earth." While perhaps not as grand as the program book claimed, the stock exhibits fully demonstrated Missouri's "worth as a state," said the *Twentieth Century Farmer*, as it praised the exhibits of cattle, swine, sheep, and horses. Beef cattle, "the magnificent

Aberdeen Angus cattle were a popular breed, according to the 1901 *Premium List*. (M. S. F.)

Below: A Jersey cow was a favored milk cow during the early 1900s, according to an illustration found in the Missouri State Fair archives. (M. S. A.)

Missouri herds, that for years have been champions of the world, will be on hand in holiday attire" to compete for $2,369.50 worth of premiums. Each animal was to be accompanied by "a competent, courteous, and neatly dressed attendant" who was to remain with the animal to answer questions asked by visitors. The attendants were also to "keep stalls in a neat and cleanly condition and space over which visitors will pass examining stock free from water and rubbish of any character." Superintendents gave admission passes to the livestock attendants who cared for the stock at the fair.

Livestock sales produced over 55 percent of Missouri's farm income at the turn of the century, according to historians Lawrence Christensen and Gary Kremer. In 1899, Missourians owned three million head of cattle, about fifty thousand of them pedigreed. Shorthorns, the most popular breed because of their versatility as both beef and dairy animals and their docile temperament,

constituted half of the purebred cattle. Herefords accounted for eighteen thousand head, and Aberdeen Angus for three thousand head. The cattle department offered premiums for individual animals as well as small "herds" consisting of five animals in four classes—Shorthorn, Hereford, Aberdeen Angus, and Red Galloway.

Missouri dairy farmers favored Jersey cattle because of the high butterfat content of their milk, though Holstein-Friesians were becoming a more popular breed because they produced more milk. Missouri's dairy industry was not well developed in 1901; however, the program pamphlet promised, "The Dairy Industry will exhibit its golden products unsurpassed in flavor and texture." The fair's dairy cattle exhibit included classes for Jerseys and for Holstein-Friesians competing for $525 in premiums. The dairy cattle exhibit was accompanied by an exhibit of dairy products, including creamery butter, dairy butter, cream cheddar and American cheeses, which were awarded $151 in prizes. In addition, the Dairy Department invited manufacturers to exhibit various dairy implements.

Missouri ranked third in the nation in poultry production; the program pamphlet claimed Missouri produced "poultry and poultry products with an annual output of $14,000,000." Almost every farm had some chickens, ducks, and geese, often cared for by the farm's women, and raising fancy poultry was a popular hobby. The program acknowledged the role women played in the poultry industry, praising "The Missouri hen, under the management of Missouri women . . . and the Missouri Poultry Breeders Association, leaders in the industry."

The poultry exhibit was the most diverse, and according to the *Sedalia Democrat*, the "best exhibit of poultry ever seen in the state." White barred Plymouth Rock, a good meat and egg producer, was the most popular breed of chicken raised in Missouri. The program pamphlet praised the "larger birds and better specimens with finer plumage, and more of them," that would be exhibited. Poultry

A prize-winning chicken was shown in the 1901 *Premium List*. (M. S. F.)

Below: *The Ladies Poultry Journal* of January 1908 recognized the contributions of women to Missouri's poultry industry. (T. P.)

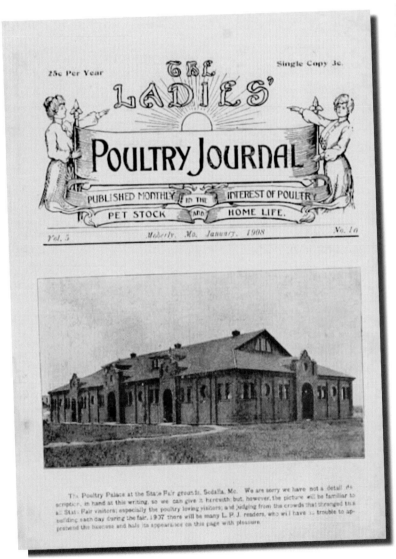

fanciers surprised Poultry Superintendent Codding and brought forty-six varieties of chickens, including fancy birds such as "frizzles" and "silkies." They also brought eight varieties of ducks, two varieties of turkeys, five of geese, and assorted pigeons. The *Premium List* acknowledged that "exhibitors will be required to furnish attractive and convenient coops" because the fair was not yet "thoroughly

equipped." Poultry breeders earned a total of $475.50 in prizes. At a time when chicken was, according to *American Heritage* writer John Steele Gordon, a "luxury that only the rich could regularly afford," the *Sedalia Democrat* described the poultry stores' display of "elegant" chicken products "arranged in an artistic manner . . . whets the appetite and inspires covetousness." The poultry stores prizes totaled $204, surely an encouragement for Missouri poultry breeders to increase their efforts to produce better chickens.

Missouri produced at the turn of the century seven percent of the hogs in the United States. Poland China hogs were the most important breed, but Duroc hogs were becoming more popular, as were Berkshires. The regulations for the Swine Department outlined in the *Premium List* focused on breeding quality, stating, "Both boars and sows, two years old and over, must have

Above: Sunnyside Hog Farm showed its prize Berkshires in the 1901 *Premium List.* (M. S. F.)

Monsees' Limestone Valley Farm exhibited prize-winning mules at the first fair. (S. P. L.)

THE MISSOURI STATE FAIR

Nature's Loom

Clothes the Sheep and It
In Turn Clothes Mankind

The Annual Sheep Show at the Missouri State Fair shows what Missouri farmers can do in breeding this useful animal that requires so little from man's hand in the way of food

**If You Have Sheep to Show Bring Them to Sedalia
Do Not Fail to See the Sheep Show**

MISSOURI STATE FAIR September 22 to 29

An early *Premium List* advertises the sheep show.
(M. S. F.)

Right: Mrs. Helen Gallie Steele, assistant superintendent of the Art Department in 1901, was active in local music and art activities.
(S. P. L.)

proved themselves breeders before the animal is entitled to contend for a premium." Hog breeders showed Poland China, Duroc, Chester White, and Berkshire swine for a total of $734 in prize money. The fair's exhibit of swine included the "pick of the best herds in this country," some of which had won blue ribbons at the Nebraska and Wisconsin state fairs.

Missouri was the most important producer of mules in the nation; in 1900, there were 244,521 mules in Missouri. The state, said the program pamphlet, had also become well known for its production of saddle horses, owning "the champion saddle stallions of the world" and achieving "prominence in breeding these equine beauties." At the fair, exhibition horses were grouped according to size and use. Heavy horses, such as Clydesdale, English Shire, Percheron, and grade draft horses, and jacks, jennets, and mules were under the charge of Superintendent Arthur Eshbaugh of Festus, Missouri. Prize money in the amount of $316 was awarded to the heavy horses and mules. Light horses, including roadsters, teams, family mares or geldings, and saddle horses, were under charge of Superintendent Charles

Busch of Washington, Missouri. They competed for $677 in prizes.

While sheep production was not as important to Missouri agriculture as cattle and hogs, Christensen and Kremer note that in 1900 Missouri farmers owned one million sheep, mostly Merino, Cotswold, Rambouillet, and Shropshire breeds. The fair's *Premium List* specifically excluded any "stubble shorn sheep" and required that the date of shearing be included with the animal's registration. The Sheep Department, under the direction of Superintendent C. J. Cloyd of Fayette, awarded prizes totaling $556 for Cotswold, Leicester and Lincoln, Southdown, Oxford, Shropshire, French Merino, and American Merino sheep, as

well as for fleeces and wools. "The admirer of livestock," wrote the *Sedalia Democrat*, "will thoroughly enjoy the fine exhibition in the sheep building."

Other natural products of Missouri were displayed in the Minerals and Forestry exhibit. Missouri was known for its mines and quarries, which the program pamphlet bragged held "practically inexhaustible" supplies of "superior quality marble, onyx, granite, and building stone." Although Missouri's mineral output did not meet the pamphlet's claim to "equal that of the Klondyke," it generated $22 million in 1900. The *Premium List* considered a "creditable display" of "commercial quantities" of lumber, building stone or minerals to be "of great importance," and suggested that counties make exhibits of their local products. Carthage and Warrensburg quarries won premiums for the best displays of building stone. Displays of Missouri wood, a product generating slightly over $11 million in 1900, were "extensive." Gallie Lumber Company of Sedalia showed an oddity, an elm tree, forked as a young tree, whose fork had later grown together leaving a hole large enough for a person to jump through. George H. Shepherd, a LaMonte nursery owner, and Lon Luther, editor of the *LaMonte Record*, mounted an exhibit of various kinds of Missouri wood. That only four categories merited $110 in prize money underscores the importance the state and the fair placed on Missouri's rocks and trees.

The arts were considered important at the turn of the century, for they "uplifted" the mind and spirits. Their importance is reflected in the $341.25 designated for prizes. The Art Display was under control of Superintendent W. B. Mackey and Assistant Superintendent Mrs. W. D. Steele, who had "skillfully arranged the exhibits." Many entries came from New Jersey and Ohio, but the *Sedalia Democrat* bragged that these were "outclassed" by Missouri artists.

A cattle barn hosted the arts and textile exhibits at the first fair. (S. P. L.)

Visual art included pictures of flowers, seascapes, fruits, animal fish or birds, figure drawings, portraits, and fancy heads executed in charcoal, watercolor, pastel, and pencil.

China painting was a particularly popular art form for women. Ladies painted flowers, animals, and fruits on plates, punch bowls, bon-bon, nut and salad dishes, fruit and bread plates, trays, pitchers, lamps, and toilet sets, powder boxes, hair receivers, hat pin holders, pin trays, soap dishes, wash bowls, pitchers, and slop jars. Pyrography or wood burning, another popular craft practiced by women, decorated stands, tabletops, picture frames, portraits, and boxes. Tooled leatherwork formed part of the art exhibit, as did fancy penmanship.

The textile exhibit, sharing space in a cattle barn with the art exhibit, allowed women to view others' work and show off their own. At a time when women had few commercial opportunities, they often demonstrated their artistic talent by knitting, crocheting, sewing, and embroidering decorative and practical items. "What the Missouri women can do in the way of providing delicacies for the table and household ornamentation will be well worth a visit to the fair," claimed the program pamphlet. The *Premium List* identifies ninety-nine separate classes of textiles and needlework. Practical needlework such as handmade rugs, portieres, clothing and aprons, knitted and crocheted shawls, afghans, and bedspreads, quilts, lap robes, carriage robes and comforters of silk, wool, and cotton, and handkerchiefs with homemade lace, along with decorative items such as doilies, dresser scarves, pin cushions in silk, cotton, and linen, and embroidered pictures, piano covers, pillow cases, and table cloths, competed for $272.25 in prizes.

Women also shone in the displays of pantry stores and baked goods. Superintendent W. L. Nelson of Bunceton, Assistant Superintendent Mrs. Anna Yeater, and Judges Mrs. Betty Gentry, Mrs. John Walsmsley, and Mrs. Frank Baird sampled four types of bread, rolls, beaten biscuits, and doughnuts. They tasted twelve different types of cakes, teacakes, gingerbread, and gingersnaps.

At the 1901 fair, Hill's Business College erected a tent and exhibited students' skills; by 1906, they had built a small frame building near the Agriculture Building. (S. A. C. C.)

They evaluated jellies made from fourteen different fruits, considering the clarity, texture, and taste. They sampled four varieties of preserves, three fruit butters, and three jams, in addition to eating a variety of pickles and relishes. After making the difficult decisions about which products were the best in each of the 103 classes, the judges awarded a total of $323 in prizes. A special prize of five dollars was awarded for the best display of two types of bread, three different cakes, and two varieties of cookies produced by a girl under fourteen.

Colleges exhibited their best students as examples of their teaching excellence. Central Business College of Sedalia featured demonstrations of touch-typing, of artistic penmanship by Professor Rogers, of dictation taken at 126 words per minute, and of rapid calculation by eight-year-old Harry Millaway. Hill's Business College, also of Sedalia, demonstrated Gregg Shorthand taken at 180 words per minute by a student watching his work and at 120 words per minute by a blindfolded student. Mollie Sharp touch-typed at 116 words per minute, an amazing speed on the manual typewriters of the day. Jones Business College Professor S. N. Falder won first prize in penmanship.

The fair's primary purpose—education—blurred into its secondary purpose—entertainment—at the racetrack. Bicycle races took place Tuesday, Wednesday, Thursday, and Friday, with members of the St. Louis Bicycle Club winning most. Automobile races included Monday's special show of automobiles powered by gas, electricity, steam, naphtha, and gasoline, as well as Wednesday and Friday's "fancy automobile track exhibition." Tuesday, Wednesday, and Thursday featured mile, half-mile, and two-mile automobile races.

The favored races, the ones that drew most visitors to the fair, were the horse races. Their special importance is seen in their position as first in the *Premium List.* Speed Superintendent Abraham C. Dingle of Moberly, Assistant Superintendent C. L. Hounsom of Glenwood, and Starter H. E. Woods of

The speed horse barns near the racetrack stabled racehorses. (V. B.)

Visitors viewed races from a covered amphitheater and open grandstand. (S. P. L.)

Norborne, supervised the races. The program pamphlet promised "Kings and Queens of the Turf" the opportunity to test their "Speed and Endurance" on the "best regulation mile track in the state" for a total of "$6500 in purses." Only one race, the 2:45 trot, was held on Monday. The track, "one of the fastest in the Midwest," was in fine condition. The race schedule announced in the *Premium List* and the races reported in the *Sedalia Democrat* differ; some rearranging of the schedule was necessary because of the weather. On Tuesday, the races were cancelled because of rain. Wednesday's races, a 2:33 pace with a $500 purse, a 2:23 trot, and a race with a $1,000 purse were finished in blinding rain. Thursday offered the best racing. The mile and one-eighth Missouri Derby, intended, according to the *Premium List* to "stimulate the breeding of the thoroughbred in Missouri," had originally been scheduled for Wednesday, but was postponed until Thursday because of rain. The fair intended to make this race a "permanent feature." One race held in conjunction with the Missouri Derby, the Derby Day Trot, featured twenty-two horses competing for the $1,000 purse. In addition, Thursday held a three-year-old pace with a purse of $500 and a 2:30 trot with a $1,000 purse and a three-quarter mile dash. Friday's races included a pace and a trot with prizes of $500 and a running race of seven furlongs or seven-eighths of a mile with a prize of $100. The higher premiums awarded at the races—nearly a third of the fair's total prize package of $20,000—indicate the popularity of racing, as well as the entrance fees racehorse owners had to pay. However, to counteract questions from the anti-gambling lobby, even the races were described as educational.

Despite the betting that took place at the races, the state fair was considered to be a "moral institution." At a time when ministers in Illinois were protesting their fair, Sedalia's ministers' association endorsed the state fair because it was closed on Sunday, and according to the General Rules and

Regulations, did not allow alcoholic beverages, "questionable or immoral shows" or "any games of chance or gambling devices" on the grounds. The Grounds Superintendent had to "investigate and report to the secretary any immoral exhibition, or the conducting of any game of chance or the sale of intoxicating liquors"; and he had the power to order "the police to make an immediate arrest" of those who violated the regulations. Most concessionaires complied, but one enterprising vendor set up a beer stand directly across the road from the fairgrounds. Sedalia Police quickly shut him down

The evening of September 13, the first Missouri State Fair closed. Competitors began removing their exhibits that afternoon at four. Secretary Rippey continued to pay prize money to winners of the various competitions. Visitors had, for the most part, been pleased with the buildings and exhibits. Some criticism apparently existed, for the *St. Louis Republic* praised the fair and denounced those who maintained a "carping spirit of criticism." The *Sedalia Democrat* pointed out that "the attendance . . . was sufficient to show great popular interest in the fair, and to show also that it will be of immense benefit to the state and to this community." Elected officials agreed. County Recorder Ewing reported to the *Nevada Mail* that the "stock shows, especially of cattle, were the finest he ever saw." State Senator Jesse Jewell told the *Kansas City Journal*, "altogether the fair is a great credit to the state." Livestock journals agreed the fair had been a success. The *Breeder's Gazette* announced, "The beginning is altogether satis-
factory and pleasing."

Visitors line the track to watch the horses compete. (D. C.)

The World's Fair Connection

Horticul
Pan.-Pac. Int. F

The Missouri State Fair has a long connection to the various World's Fairs. During the first three fairs, dignitaries visited for World's Fair Day in celebration of the St. Louis World's Fair. Later fairs expanded the world's fair connection as Missouri exhibitors took prize-winning animals and painting from Missouri to the World's Fairs in San Francisco and Chicago. The Missouri Building, so named in 1935, hosted the Missouri display from the Chicago Fair of 1933 and 1934.

The 1915 Panama-Pacific International Exposition in San Francisco used the talents of Missourian C. W. Steinman. The head of Steinman Orchards in Dalton, Missouri, and a circuit-riding minister in the Methodist Church, Steinman volunteered to help landscape the Missouri State Fairgrounds. In San Francisco, Steinman arranged an "array of apples, nuts, vegetables, watermelons and flowers, beautifully displayed under green latticed archways." Laura Ingalls Wilder, writing for the *Missouri Ruralist*, praised the Grand Medal of Honor winning exhibit: Missouri went to the fair to show . . . what she could do in agriculture at home and she did it."

W. C. Steinman, fairgrounds volunteer, designed this exhibit for the 1915 Panama Pacific Exposition. (Photo courtesy Sandra Blunk)

Another World's Fair connection happened in 1939 at the New York Fair. The "Romantic Missouri" display, with its large murals painted by Missouri artists, exhibits of native rock and timber, replicas of log cabins, maps, and booths showing Missouri products was built at the fairgrounds. F. W. S. Sayers of Jefferson City directed the exhibit. He supervised its creation, orchestrated the over one hundred artists and workers who built the display, and accompanied it and a work crew to New York to install it at the World's Fair.

...xhibit of Missouri,
...tion, San Francisco, 1915.

CHAPTER 3

A STUDY IN STYLES:

Architecture and Design

The Horticulture Building, built in 1903, is now the Varied Industries Building. (M. S. A.)

PRIOR TO THE FIRST FAIR, VARIOUS LIVESTOCK BREEDERS URGED the General Assembly to make generous appropriations for exhibition buildings on the state fairgrounds. The Missouri Shorthorn Breeders Association, for example, asked for enough money to "provide equipment commensurate with the great industrial interests of the state." However, the state only provided $50,000, enough to build a racetrack, stables, and five livestock barns. As a result of the limited funds, at the first fair two of the cattle barns had to be used for other displays, and many of the exhibits were housed in tents. After the 1901 fair, the *Breeder's Gazette* praised the existing buildings, saying, "Whatever has been accomplished has been well done" and indicating that the existing facilities provided a good foundation for further growth. However, the *Gazette* suggested that Missouri "must continue

suitable appropriations for the building fund until these grounds will compare favorably with those of other states even less great agriculturally than Missouri." In early 1903, the "Preliminary Statement" of The *Missouri State Fair Premium Book* noted, "Permanent buildings, especially designed for the purposes intended, should be considered and appropriations made for their construction as speedily as the revenue of the state will warrant." As the development of the fairgrounds continued, attitudes toward exposition and fair design, prevailing architectural and landscaping styles, and changes in agriculture influenced the planning and building of the Missouri State Fairgrounds.

Buildings have been erected throughout the years of the fair, but three major time periods showed significant construction. The first followed the establishment of the fair and lasted approximately fifteen years. During the Depression, the Works Progress Administration assisted the fairgrounds and erected another group of buildings and created additional landscaping. A third emphasis on building happened during the 1960s, 1970s, and 1980s. Recent design activities have focused on implementing a Master Plan for

The first Machinery Building is shown on a postcard. (S. P. L.)

A plat map shows the fairgrounds in 1915. (M. S. F.)

building and renovation established in 1998, on rehabilitating buildings to make them both more compatible with Bast's original plan, and on making the buildings and grounds more accessible and appropriate for contemporary fairgoers.

Fair historian Wayne Campbell Neely points out that in the early twentieth century, fairs began to construct permanent sites with brick and steel buildings. By establishing its state fair when it did, Missouri

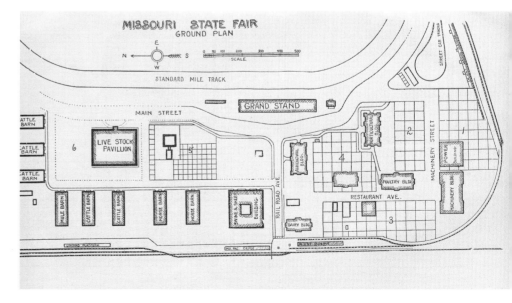

The first Administration Building overlooked the racetrack; the grandstand is visible in the background. (S. A. C. C.)

Below: The original horse barns housed exhibition horses and mules. (S. P. L.)

avoided a problem many other states faced during the late nineteenth century—that of temporary buildings dismantled and moved from site to site each year. Missouri began with a permanent site and a plan for its development, influenced by attitudes toward the purpose of fairs, as well as by popular exposition architecture and landscape design.

The 1893 Columbian Exposition, commonly known as the Chicago World's Fair, influenced fair design by encouraging the logical arrangement of elements at fairgrounds. Historian Halsey Ives points out that the prevailing attitudes at the time of the Chicago Fair encouraged the development of a very organized design in which buildings were placed according to their functions, and entertainment was separated from the educational exhibits. Geographer Fred Kniffen, in describing the pattern of state fairs at the turn of the century, suggests that the "educational . . . displays of livestock, crops, and the fruits of home arts and crafts" remained the most important element of a state fair, but that by the early twentieth century, "sports and amusements threatened more than ever to run away with the fair." The racetrack became a prominent feature of most fairgrounds, vying with exhibition spaces for importance, and space was provided for games and shows. This pattern of development is apparent in the original site plan for the Missouri State Fair, which shows the racetrack as the central feature of the grounds, with stables and the Administration Building close to the track and exhibition buildings somewhat separated from the track.

In addition, the Columbian Exposition encouraged the development of a unified architectural style for fairgrounds buildings. Thomas Bast's plans for the buildings on the Missouri State Fairgrounds for the most part reflected his signature eclectic style, which combined elements of several architectural styles—Romanesque Revival style, the Mission style, and the Arts and Crafts

style. The Romanesque Revival, popular from 1880–1900, with its round arches, columns, and towers adorning massive masonry buildings, was widely used in Sedalia. The Mission style, which developed in the 1890s, used round arches and square towers, while incorporating Spanish and Moorish influences such as shaped domes, decorative parapets, and finials. The Arts and Crafts style emphasized a more horizontal line, with open rafter ends and stick work ornamentation. Bast borrowed the round arches and massed masonry from the Romanesque Revival era, but used the shaped domes and parapets of

Paving contractor George L. Cooley of Westlake, Ohio, paved several hard-surfaced roads on and around the fairgrounds in 1906. (S. A. C. C.)

the Mission style for the red brick livestock and exhibition buildings. He identified these styles in his drawings as "Spanish Renaissance" and "Renaissance." He used Arts and Crafts style stick work for the wooden barns and buildings. Though similar in style, each Bast design maintains its individuality because he used a variety of decorative elements such as windows and parapets. As such, the buildings become a showcase of Bast's abilities as an architect.

Bast's plans for individual buildings reflected the typical exposition hall style of the time. Since the directors had visited fairs in other states, they were aware of the type of buildings generally used. Bast had in his files photographs of the fairgrounds in Columbus, Ohio, and Springfield, Illinois. State fair buildings in Ohio and Illinois, as well as those in Iowa and Indiana, resembled in many ways public buildings such as courthouses. They generally featured multiple entries, at least one on each side of the building. Entrances were indicated by porches, by extensions called pavilions, or by decorative embellishments such as pilasters or columns. The multiple entrances often enhanced a cruciform plan of exhibits lining the walls and forming a cross shape in the center. Central domes with ventilating mechanisms emphasized the cruciform plan. Buildings often had corner towers accented with turrets topped with decorative finials. Livestock buildings were designed for practical use. Barns generally had corridors marked by decorative entrances, with stalls opening off the corridors. Larger livestock buildings often resembled other exposition buildings and featured amphitheater-style show rings in the center for judging and showing animals.

The Columbian Exposition also influenced the development of the "city beautiful" movement, which emphasized the artistic design of landscapes with broad, interconnecting streets, decorative plantings, and public art. George Kessler of Kansas City, a prominent landscape architect, and his employee Henry Wright had drawn plans for the fairgrounds and outlined roads, walkways, and areas of plantings. As a result of Kessler and Wright's work, the Missouri State Fair was laid out in a logical fashion with walkways, trees, and flowerbeds that accented the distinctive style of its buildings. In the fall of 1901, fair custodian W. S. Allison planted more than one thousand trees, interspersing quick growing Carolina poplars and soft maples with slower growing and more long-lived walnut and elm trees. The State Fair Board directed the custodian in 1902 to "dig forest trees" such as elm, sycamore, ash, sugar trees, and soft maples and continue planting trees on the grounds. During the spring of 1903, Kessler supervised additional plantings. The city of Sedalia added to the effect Kessler and Wright planned by creating boulevards lined with trees and flower beds leading to the fairgrounds. Third Street widened into a boulevard at Park Avenue. The boulevard continued west to intersect with State Fair Boulevard, which led from Third Street south to the Main Gate of the fairgrounds. Sixteenth Street widened into a boulevard at Barrett Avenue and continued west to intersect with State Fair Boulevard at the fairgrounds' Main Gate.

The layout and buildings also reflected the interests and emphasis of agriculture at the time. Breeder's Fund money, although designated for the overall development of the fairgrounds, needed to be spent on racing to avoid conflict with powerful racing interests and St. Louis racing clubs. Thus, the racetrack was begun first, in 1900, on a site selected by surveyor T. O. Stanley. Completing the track eventually cost $20,000, a significant portion of the money allocated for the fairgrounds from the Breeder's Fund and the General Assembly appropriation. Stables for racehorses followed the racetrack. Other 1901 buildings reflecting the state's agricultural background included frame show horse barns, cattle barns, and a swine and sheep pavilion.

The ten speed horse barns were one story, with stalls opening onto the outside. Each had

The racetrack and a cattle barn are shown in this 1906 photo. (S. A. C. C.)

a ventilation structure on the roof. The exhibition horse barn was a frame building with a two-story ventilation structure on the roof. Stalls opened off a central aisle. Corner towers accented each end of the building and two hipped-roof gables adorned each long side. The first cattle barns featured corner turrets and ventilation structures extending almost the length of the buildings. Windows in the central gables and the corner towers provided additional ventilation. The 1901 swine and sheep pavilion boasted rounded entrances and four corner towers. A central ventilation structure rose above a row of ventilating windows.

The original Administration Building was a frame building resembling a home in the foursquare style popular for domestic architecture at the time. It was a two-story square frame building with a porch supported by round columns on the east façade and a balcony over the porch. Decorative brackets supported the roof's overhang. The building housed offices and an emergency hospital. This building, immediately south of the grandstand, was close enough to the racetrack that dignitaries could view races from the balcony.

The 1903 "Preliminary Statement" suggested that the existing barns were not large enough for the livestock displayed, and proposed a building program which would include poultry and dairy buildings, an amphitheater for the display of livestock, an implement hall, a building in which to display manufacturer's exhibits, and a building in which to exhibit minerals, as well as buildings for textiles, arts, and horticulture, so that the cattle barns could be used for housing and showing livestock.

The state continued to add to the buildings on the fairgrounds following Bast and Kessler's plans. In early 1903, the fair added three buildings. The buildings were of brick and steel, reflecting a desire for permanence and fire protection, and are still in use. In response to pressure from poultry breeders, who constituted a fourteen million dollar industry in 1900, the state contracted Joseph Heckert and S. Wilson Ricketts to build a brick and steel

The original cattle barns in 1903. (S. P. L.)

The Poultry Building served many purposes. (T. P.)

PUBLISHED BY R. A. DUNLAP, SEDALIA, MO.

8356 Poultry Building, State Fair, Sedalia, Mo.

George Kessler and Henry Wright

The "city beautiful" movement of the early twentieth century recognized the importance of attractive urban surroundings and beautiful parks. Two of the leaders of the city planning concept—George Kessler and Henry Wright—worked on the landscape plans for the Missouri State Fair.

Kessler, born in 1862 in Frankenhausen, Germany, came to the United States in 1865. He grew up in Hoboken, New Jersey, and Dallas, Texas, and returned to Germany in 1878 after his father's death. He studied botany, landscape gardening, and engineering, and studied the civic design of Europe's major cities. He returned to the United States in 1882, and worked with Frederick Law Olmstead on New York City's Central Park. In 1890, Kansas City hired Kessler to design plans for Kansas City streets that utilized what he called the city's "eccentric topography" of rivers, bluffs, creeks, and hills. Kessler worked in twenty-three states as well as in Mexico and China.

Henry Wright, born in 1878 in Lawrence, Kansas, attended the University of Pennsylvania. Following graduation, he moved to St. Louis. In 1901, he was employed as an assistant to George Kessler. During this time, the two designed the Missouri State Fairgrounds. Wright would also design the west and central wings of Bothwell Lodge, home of John Bothwell.

Both Kessler and Wright saw the importance of natural areas or parks in urban areas. Parks enabled city dwellers to experience nature and to enjoy trees and green spaces. Their design for the Missouri State Fairgrounds provided a pleasant, shady area in which to view the exhibition. Their suggestion that wide boulevards planted with trees and flowers lead to the fairgrounds enhanced the city of Sedalia as well as the fairgrounds.

The *Journal of the American Institute of Architects* eulogized Kessler as "a city planner and landscape architect of extraordinary insight, vision, and practical ability."

Poultry Palace 55 by 122 feet. Each façade features projecting pavilions highlighted by double doors; flagpoles on the roof mark the pavilions. Decorative terra cotta accents the round brickwork arches. A long rectangular ventilation structure on the roof, now removed, provided fresh air to the building. The building was dedicated during the fair at a poultry show described by the *Sedalia Democrat* as the "best ever held in the West." In 1905, this poultry building became the Dairy Building; it became the University Building in 1920, and became the Future Farmers of America Building in 1955.

The fair directors also arranged for Thomas Johnson to build a Palace of Agriculture, 60 by 160 feet, for the display of farm products. Bast's design identifies the style as "Spanish Renaissance." The $22,800 building features shaped parapets above each of the four entrances, a hipped roof, and corner towers with hipped roofed dormers originally topped with galvanized iron finials and urns. The towers feature rounded windows resembling those in bell towers, adding to the Missionesque style of the building. The 1906 *Premium List* labels a photograph of this building the Horticultural Building. Though postcards from 1908 through 1912 call this building the Liberal Arts Building or the Textile and Fine Arts Building, the *Program Books* of 1915, 1918, and 1920 identify this building as the Agricultural Building. In the 1920s, it became the Education Building, showcasing the work of Missouri's elementary and secondary schools. It is now the Commercial Building.

Heckert and Ricketts, with help from Sedalia masonry contractor John Colaflower and the Dean Brothers Construction Company, built the Horticulture Building, perhaps the most decorative of the 1903 buildings. Bast called the design "Renaissance." The building has a somewhat Moorish look, featuring rounded domes accented with galvanized iron urns topping corner towers with round windows. The east and west entrances are highlighted by shaped parapets and are recessed within arcaded pavilions. Towers topped with octagonal turrets and domes flank the entrances on the east and west facades of the building. Flagpoles originally topped the turrets and domes. The 86-by-168-foot building with a hipped roof and six dormers cost $24,794. The 1906 *Premium List* identifies this as the Agricultural Building, as do post-

The Agriculture Building was also called the Horticulture Building. (T. P.)

Horse races provided popular entertainment for visitors in the Grandstand. (M. S. F.)

cards from 1909 through 1912. However, the *Program Books* of 1915, 1918, and 1920 identify it as the Horticultural Building. After World War I, it became the Varied Industries Building.

In September 1903, a fire destroyed many of the frame cattle barns on the grounds. The Executive Committee of the Board of Directors decided to replace the destroyed buildings and to extend the steel grandstand by 160 feet, but again had to consider costs. The Breeders' Fund contained $2,082 and was expected to earn $3,000 more during the coming year; the surplus left over from the 1903 state fair was $1,600. The board expected Governor Dockery to release the funds he had withheld from the $25,000 appropriated by the legislature. The grandstand was completed, and the *Nebraska Farmer* lauded it as "unequalled from Chicago to San Francisco."

In 1904, Johnson, following Bast's designs, built the Shorthorn Barn, 65 by 160 feet, with two round arched entrances flanked with pilasters. Its parapet of decorative brickwork adds to the Missionesque nature of the building, as did the original tile roof. In 1904, Johnson also built a heavy horse barn and a light horse barn to replace those destroyed by the fire. The horse barns were each 72 by 164 feet, of brick and steel with round arched entrances, hipped roofs, and pilasters. Decorative details are different, however. Both horse barns' pilasters, capped with metal pyramids, extend above the roofline. The light horse barn has a projecting bay that provides office space.

Bast designed two cattle barns in 1905. Cattle Barn Number 2, the Hereford Barn, was the more decorative, featuring pilasters marking the round arched entrances and the corners. Gable ends and cross gable ends feature tripartite round arched windows. Other windows have segmental arches. The other barn, which now houses Guernsey cattle, has slightly corbelled flat arches on the end facades, and gable ends with Palladian windows with hoodmolds.

By 1905, the original poultry building had been outgrown, and a new Poultry Building was built to replace it. The 80-by-160-foot building features

central entrance bays, round arched windows, pilasters, and terra cotta trimmed parapets. Like many other buildings, it changed its name and function, becoming the Missouri Building in 1935 when it housed the Missouri state exhibit from the Chicago Century of Progress Exposition. It became the 4-H building in 1958.

Agricultural journals praised the rapid development of the fair's facilities. According to the *Nebraska Farmer*, "The Missouri State Fair has done what no other State Fair in the United States has done in the same number of years. She has built in four years more buildings of substantial structure and beautiful architecture than has ever before been accomplished by any other State Agricultural Association." The *Twentieth Century Farmer* agreed, calling the Missouri State Fair "unequaled in the substantial character of its equipment and in the architectural beauty of its buildings."

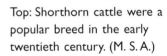

At the 1906 fair, Governor Joseph Folk dedicated the Coliseum, the largest of the early buildings and still a centerpiece of the fairgrounds. The 180-

Top: Shorthorn cattle were a popular breed in the early twentieth century. (M. S. A.)

A 1915 postcard shows several cattle barns. (M. S. A.)

Left: A new Poultry Building was built in 1905. (M. S. A.)

by-235-foot building, with its surrounding arcade, repeats the arch motifs, terra cotta trim, and pilasters of the brick barns, but adds classical elements that had become popular following the St. Louis World's Fair The pedimented portico is supported by square Doric pilasters, and is accented by a circular window and dentil molding. The four corners and the portico feature two-story rounded archways highlighted with decorative brickwork in a series of elaborate soldier courses. The five dormers feature tripartite-arched

The Coliseum is a center-piece of the Missouri State Fairgrounds. (P. C. H. S.)

Right: In 1907, T. H. Johnson built two more cattle barns, housing Jersey and Angus cattle. (M. S. A.)

Thomas Bast, State Fair Architect

Thomas Bast, one of Sedalia's "clever and progressive young business men," typified the proper businessman at the turn of the century. Sedalia's most prolific architect, he designed the fairgrounds as well as other Sedalia landmarks including the east wing of Bothwell Lodge.

Bast was born in Wright City, Missouri, on October 16, 1863. His father, David A. Bast, was a carpenter, contractor, and builder. Thomas Bast lived with his family in Wright City for eighteen years, attending the public schools there and learning carpentry from his father. In 1881, Bast moved to Texas, working as a carpenter in Dallas and at Graham. In 1882, Bast moved to St. Louis where he worked as a contractor during the day and studied architecture at night at the School of Art and Design of the Polytechnic Department of Washington University.

Bast moved to Sedalia in 1889 upon completion of his studies. Within five years of his arrival here, he had designed 150 buildings in a variety of styles. Bast's public buildings included Bothwell Hospital and Washington School, Mark Twain School, Whittier School, and with the William B. Ittner Company of St. Louis, Smith-Cotton High School. The public buildings are utilitarian in design and well built; most remain in use. He designed the Gothic styled United Church of Christ, also still in use. He supervised the design and the later remodeling of the Citizens National Bank building at Main Street and Ohio Avenue in the Beaux Arts Classic style. The home at 800 West Seventh, a typical Victorian, was a Bast design, as probably were the homes at 1108 West Broadway and 1012 West Broadway. His largest and most lasting legacy is the design of the Missouri State Fairgrounds, on which he used a stylistically eclectic mix of Romanesque Revival, Mission, Arts and Crafts, and Classical Revival styles.

Bast married Katie Steele in 1888. The couple had five children—Steele, Mary, Tom, Jr., Sam, and Joe. A member of the Methodist Episcopal Church South, he served as steward, and was Past Prelate in the fraternal order the Knights of Pythias. He was also a member of the Sedalia Lodge No. 1, Royal Tribe of Joseph, and the Modern Woodmen. A Democrat, he was involved in local politics; following his retirement from his architectural practice, he served as city treasurer.

In 1895, Bast had predicted a major building boom for Sedalia. Before his retirement in 1929, he helped make it happen.

windows, as do the gable ends. The interior has amphitheatre seating around an oval arena used for showing cattle and horses.

In 1907, Johnson again erected buildings on the fairgrounds. The Jack and Jennet Barn features a stepped parapet on its front and rear facades. The windows echo the stair step motif, as does the decorative brickwork between the pilasters. The Cattle Barn East, now the Jersey Barn, has circular windows with terra cotta keystones and a circular parapet. Its projecting central bay houses a small office. The Cattle Barn West also has a projecting bay for an office. It features a tripartite window in the gable accented with molding whose lines are repeated in the parapet. The arched entrances are accented with three rows of header bricks. It is now the Angus Barn.

In 1909, the John Deere Plow Company funded the erection of a building to showcase its farm implements. The building, simpler than other fairgrounds buildings, was originally open on the sides. The four projecting entrances are marked with shaped parapets. After the John Deere Company stopped using the building, Pettis County used the building for its exhibits, and attempted unsuccessfully to purchase the building. During the 1920s, the building was used for the State Fair Kennel Show. In 1925, the State Fair purchased the building for $16,500 and enclosed the side openings with windows. Home Economics moved into this building in the late 1920s.

Bast designed another machinery building in 1909. It was a square 120-by-120-foot,

The John Deere Plow Company exhibited gasoline motors in 1909. (D. C.)

Bottom: The 1909 Machinery Building was open on sides and top. (T. P.)

Machinery Building, State Fair, Sedalia, Mo.

one-story building, originally with a canvas awning that served as its roof. Each façade had six large openings with stone lug sills. As farming became more mechanized, the machinery exhibits were moved to an outdoor area on the south and west of the fairgrounds, and in the 1920s windows and doors were added, and the Machinery Building became the Missouri Building. In the 1930s, when the 1905 Poultry Building became the Missouri Building, the Machinery Building became the Poultry Building. The building shares its west wall with the Hall of Religion, which in the 1930s housed, according to a WPA description, "displays of all religions, apiary, agricultural, horticultural and county exhibits." A 1961 renovation removed the ventilation structure from the roof. The building, with an annex created out of the original Hall of Religion, continues to house poultry and rabbits.

In 1910, Bast used an entirely different style for the Womans Building. The tan brick building in the Georgian Revival style looks like an elegant home, a style thought by architectural historian Karen Grace to be appropriate for a building celebrating the accomplishments of Missouri's women. It features a pedimented portico supported with Doric columns. A porch supported by round wooden columns and two square brick columns runs the full width of the building. The Womans Building had dormitory space on the second floor and a ballroom or assembly hall on the third floor, a day nursery in the rear wing where mothers could leave their children while they visited the fair, and a playground for the children. It was used for many years for the display of fine arts and crafts, and for musical performances.

Two buildings built in 1913 continue the Missionesque style. The Fire Station, a small building 24 by 50 feet, features pilasters and a three-bay main façade. It served as a Demonstration Kitchen. The State Fair Dining Hall, which during the 1940s and early 1950s was used for Boy Scout and Girl Scout exhibits, is now the Missouri Beef House. It also has a Missionesque parapet and a three-bay main façade; the middle bay features entrance doors with transoms.

In the early years, the Main Gate of the Fairgrounds carried out the Missionesque appearance of many of the buildings. Each side of the gate featured a tower of three levels in graduated sizes; the top level of each had rounded windows in a "bell tower" style. The top level was accented with

Top: The Womans Building featured a day nursery and a dance hall. (D. C.)

Children play at the playground at the Womans Building. (M. S. F.)

urns on each corner and an octagonal tower topped with a dome-shaped roof. Spanning the towers was a lattice with a center medallion featuring the Missouri state seal and the words "Missouri State Fair." Thirteen arrows, representing the original thirteen colonies, extended from the state seal. Gates fastened at the center island, which housed a ticket booth.

In the early 1920s, the General Assembly appropriated $125,000 for a larger Swine and Sheep Pavilion. Bast's original plans called for a building with a center show ring, cross-hipped ventilation structure, arcaded areas on the east and west sides for animal pens, and corner towers with sleeping rooms. The central section and the east side of the building were completed, but the west side would not be built because the building had cost more than the state had anticipated. In the early 1960s, a metal building was added to the west side to provide space for animal display. The addition mars the symmetry of the building since it does not match the existing architecture. During the 1920s or 1930s, two firms submitted blueprints for a new Sheep Pavilion. One, by Victor DeFoe, follows the traditional designs prepared by Bast. The other, by George Holcomb, features an Art Deco façade. The proposed Sheep Pavilion was not built, and the Swine Pavilion continued to house sheep until 1965, when a metal Sheep Pavilion was built to the south.

Missouri's dairy industry had grown, and in 1926, Bast designed a barn for exhibition of Holstein cattle. Dean and Hancock built the barn, which fits between the Guernsey Barn and the Jersey Barn. It is architecturally simpler than the earlier barns, with straight windows and door openings. In 1929, architect Victor DeFoe followed the style set by Bast and designed a show horse barn. Named for Governor Henry S. Caulfield, it has three aisles with round arched entrances. Built by Dean and Hancock, the building blends easily with the surrounding buildings.

As the automobile became more widely used, the State Highway

Department sought federal highway monies and passed a bond issue to improve Missouri's roads. During the 1920s, the State Highway Department built the Highway Gardens to draw attention to its work. Shaded by trees, it resembles a roadside park, featuring attractive floral plantings, rock walls, picnic tables, benches, pools with water lilies and goldfish, and drinking fountains. An exhibition building housed Missouri exhibits.

In 1926, the State Fish and Game Commissioner Keith McCanse designed a pavilion in the rustic style built of logs from a variety of Missouri trees. In 1935 and 1936, the WPA built a new Wildlife Pavilion using Missouri materials, including pink granite. The building was remodeled in 1952, 1986, and 1995. A casting pool built in 1959 highlighted sport fishing in Missouri's many streams and rivers. Now called the Missouri Department of Conservation Wildlife Pavilion, the area displays Missouri fish, reptiles, and animals.

Left: The Highway Garden resembles a park. (M. S. A.)

Below: The Conservation Pavilion hosts an exhibit of Missouri fish. (M. D. C.)

The current Administration Building, named for Governor Sam A. Baker, replaced the existing Administration Building in 1927. It is a two-story building in the Classical Revival style built on a raised basement. Bast used a Classical Revival style with a symmetrical façade and a portico with square stone columns. Built of buff colored brick, it formerly housed bedrooms for the fair's Board of Directors and a cafeteria in its basement. The building's plumbing and electrical systems were updated in 1964. The building now houses office space for the fair's administrative staff.

One very popular area designated during the 1920s reflected changing attitudes toward fairs. A 1918 advertisement suggested that "now most everybody knows" that a Midway is "a place of varied amusements." As entertainment became a more important part of fairs and amusement parks a part of major cities, the Missouri State Fair set aside a Midway for the "entertainment for the masses."

Victor DeFoe used the Classical Revival style in some of the buildings he designed, such as the Floriculture and Art

The dedication of the Sam Baker Building highlighted the 1927 fair. (M. S. F.)

Right: A happy fairgoer finishes a Midway ride in 1941. (V. B.)

Building, a two-story frame building with a two-story pedimented portico supported by eight square columns, built in 1929. Originally displaying floriculture on the ground floor and art and antiques on the second floor, this building is now devoted exclusively to Fine Arts. In the early 1990s, the Missouri Arts Council financed a major rehabilitation of the building, refinishing walls and adding lighting to make it more suitable for the display of artwork. The addition of an elevator made the second story handicapped accessible.

During the 1930s, the Great Depression struck and drastically affected Missouri agriculture. Sedalia, particularly hard hit by bank failures and factory shutdowns, became the site of many Works Progress Administration projects. The state fairgrounds benefited, as WPA crews erected several buildings. Most of the WPA buildings were utilitarian—permanent concession stands and comfort stations (restrooms), but exhibit buildings and entertainment facilities were also built. In the early 1930s, Victor DeFoe designed a new Floriculture Building in a style similar to but less elaborate than that of the Fine Arts Building. The WPA constructed this building just east of the Fine Arts Building.

An aerial view is shown on a postcard from the early 1940s. (T. P.)

The Main Gate is an example of Art Deco design. (V. A.)

The WPA built another speed horse barn in 1937, reflecting the continued interest in racing. It was the largest barn on the grounds. In addition, a half-mile racetrack was built, centered in front of the grandstand. The WPA also built a decorative lagoon in the center of the racetrack. In addition, the WPA built a judging booth, and a stage and dressing rooms for performers, emphasizing the increased importance of shows and concerts as part of the fair's entertainment package.

In 1939, a larger and more elaborate main gate was built. The Main Gate, in an exuberant Art Deco/Art Moderne style designed by Arthur J. P. Schwarz and by Harvey King, has three ticket booths spanned by steel arch work. The Missouri State Seal highlights the arch, and Pettis County welder Paul Roberts worked the name "Missouri State Fair" in ribbon steel. In 2002, the steel arch was repaired, and the gate repainted.

In 1941, Arthur J. P. Schwarz designed the Donnell Building, or Junior Activities Barn, which housed livestock shown by 4-H and FFA members. The barn shares walls with adjacent barns and blends with the other barns stylistically, having a stair step parapet and round arched entrances. Its name appears in a pre-cast stone inset above the doors. The outbreak of World War II in late 1941 effectively stopped building on the fairgrounds. After the war, building began with the erection of more utilitarian buildings—concession stands, comfort stations, and maintenance garages.

During the 1950s and 1960s, styles and means of construction changed and the new livestock buildings changed accordingly. A wash rack used to prepare livestock for show was built between 1950 and 1960. The Donnelly Polled Hereford Barn, essentially a pole barn with a flat roof supported by metal beams and posts built in 1956, sits between the Angus and the Guernsey Barns. The milking parlor, built in 1957, was a corrugated steel building with windows that allowed visitors to watch cows being milked by machinery. Agricultural changes were apparent in the erection of barns for exotic breeds of cattle such as Charolais and Simmental. During the 1960s, a number of other metal frame buildings were added. In 1961, Norman C.

Atkins designed a rectangular metal frame building which is now the Agriculture Building. This building was the first climate controlled building on the grounds. The Consumers' Food Pavilion, built in 1967, also has a metal frame and walls of corrugated metal. In 1965 a show horse barn was built. In 1968, the Missouri State Highway Patrol built a metal building to house Otto the Talking Car.

Early in the 1950s, a wood and steel grandstand was purchased from the South Plains Fair Association in Lubbock, Texas, and moved to replace a section of the original grandstand. It was quickly outgrown, however, and in 1968 Governor Warren G. Hearnes officially opened the new grandstand. The firms of Rathert and Roth and Kenneth Balk and Associates designed the reinforced concrete structure that features a metal canopy roofing much of the seating area, and restrooms and concession areas under the stands. A separate ticket office stands partially under the south end of the grandstand.

Two buildings from the 1960s reflect increased youth participation. The Youth Building and Sales Arena, designed by Sammons and Buller and built in 1969, provides dormitory rooms and cafeteria space for approximately two hundred young people, and a show room and sales arena for animals. The Children's Barnyard, one of the most popular displays on the grounds, is a steel-framed building with a false front in the shape of a wooden gambrel-roofed barn. Located east of the Agriculture Building, it is staffed by the Future Farmers of America.

Since the fair's beginning, farm businesses and organizations highlighted their activities by building permanent buildings. Manufacturers were first

The 1961 Agriculture Building was the first climate controlled building on the fairgrounds. (M. S. F.)

Below: The Mathewson Exhibition Center houses exhibits, concerts, and off-season rodeos. (M. S. F.)

given the right to erect buildings on the fairgrounds in 1902, when the German Kali Company, a fertilizer manufacturer, applied for permission to build a building to display its merchandise. Over the years, many other companies have built buildings. The MFA Company built a feed and forage building, the U.S. Truss Steel Model Home was erected as a demonstration, and the Automated Feed Systems Company and the Farmco Company built demonstration buildings. The Farm Bureau built a one-story building in 1970 to highlight the work of its organization. The Gastineau Log House showcased the renewed popularity of log homes.

The highlight of the 1980s was construction of the Exhibition Center, named in 1994 for Senator James Mathewson of Sedalia, who encouraged its building. The modern styled, 86,000-square-foot metal building features a 22,000-square-foot arena, amphitheatre seating for 3,154 people, plus forty permanent seats for the handicapped, and an upper concourse, suitable for exhibits. It was designed by the architectural firm of Black and Veatch.

Womans Building Misspelling

The Womans Building prompts a question frequently asked by scholars, historians, and grammar buffs: Why is the name of the building spelled and punctuated as it is—incorrectly? Don Yoest and Connie Patterson of the Department of Natural Resources provide this explanation: The name "Womans Building" was imprinted on the building's portico in 1910 and has not been changed.

During the 1990s, private donors worked with the state to erect two buildings. In 1998, the MoAg Theatre was built. The 9,600-square-foot concrete building used for entertainment cost $280,000. MO-AG Industries donated $25,000 of its cost. In 1999, a 4,800-square-foot dairy center was built and named in honor of prominent Missouri dairyman and former Dairy Cattle Superintendent Bud Gerken. It contains a milking parlor and a dairy bar selling milk drinks and sandwiches. Ninety thousand dollars of the building's $262,000 cost came from contributions by the dairy industry. In 2000, the Missouri Rural Electric Cooperatives built a 9,600-square-foot building housing a media center and space for radio broadcasts. Much of the building's $950,000 cost came from contributions from Missouri's electric cooperatives. The building is especially interesting because its style harks back to Bast's designs.

Since the earliest fairs, visitors have camped during the time they spent at the fair. In 1907, the fair purchased twenty-four acres of land and designated an area between the racetrack and Limit Avenue as a campground. Because many of the tents were white canvas, the area came to be called "White City." In 1940, the fair set aside an additional forty acres for camping, and in the 1960s, established a sixty acre campground on the west side of Clarendon Road. In 1960 and 1966, restrooms were built on the campground, and in 1970, a bathhouse was built. In 1982, the shelter house added to this campground made it a popular site for off-season events. These campgrounds have water and electric hookups for recreational vehicles.

The Fairgrounds Charrette in 1995 brought together architects, historians, and fair staff from throughout the state, as well as eleven Sedalia youngsters. The participants were divided into groups and recommended renovations for the fairgrounds. Some of these suggestions have been incorporated into a Master Plan, approved in 1998 and funded in 1999 by the General Assembly. The Master Plan reflects changes in attitudes toward fairgrounds layout while at the same time preserving much of the historic nature of the Missouri State Fairgrounds. It provides for major changes in the arrangement of the fairgrounds, especially in access and parking. Some elements of the Master Plan have already been implemented. In 2001, Sedalia's Clarendon Road, which formerly bisected the fairgrounds and separated the rodeo arena from the grounds, was rerouted so as to encompass all the grounds except one camping area. In 2001, the state began construction of a new National Guard Armory adjacent to the Mathewson Exhibition Center. Built in 2002, the "Centennial Entrance" on Highway 65 draws attention to the reorientation of the grounds. A ticket plaza marks the separation between the parking areas and the fair itself. The mile racetrack has been eliminated, as it is no longer

Many visitors camp during the fair. (M. S. F.)

used. A new Future Farmers of America Building replaces the existing FFA Building, which in 2002 was used for the Missouri Frontier Exhibit, detailing the Lewis and Clark Voyage of Discovery Bicentennial. The Master Plan calls for buildings to be updated by adding heating and air conditioning so they can be used year round. The Master Plan also calls for widened walkways and additional landscaped areas for picnicking and relaxation.

When the Master Plan is fully implemented, the Missouri State Fairgrounds will be ready for its second hundred years. It will reflect the historic nature of the original grounds and provide space for the types of exhibits that will mark the twenty-first century.

Workers construct the Centennial Entrance in the summer of 2002. (V. B.)

CHAPTER 4

SHOW ME MISSOURI:
State Agencies on Display

SPEED DEPT   Police Headquarters   HOSPITAL   HIGHWAY PATROL

The Missouri State Highway Patrol used the Administration Building for its office. (S. D.)

THE MISSOURI STATE FAIR BEGAN NOT ONLY AS A SHOWCASE of Missouri agriculture, but also as a showplace for the state of Missouri. At the first fair, the University of Missouri sent a lecturer to demonstrate crops produced at the College of Agriculture. Other state colleges and universities began to exhibit products produced by their students and to advertise their academic programs. State agencies soon realized the fair provided a perfect venue for promoting their services to Missouri residents. The State Game and Fish Department, the Missouri Highway Department, the Missouri Highway Patrol, and the State Forestry Department mounted exhibits

Lieutenant Governor Long welcomes visitors at the Admission Gate. (M. S. A.)

Below: Farmers exhibit prize winning fruit and vegetables in the Agriculture Building. (M. S. F.)

and created locations for their departments. State institutions such as prisons, reformatories, and hospitals for the handicapped exhibited inmate arts and crafts.

Steve Mahfood, director of the Department of Natural Resources, acknowledged the importance of the fair when he said: "The Missouri State Fair is the biggest single public showcase of the year where we interact with people . . . In essence, our exhibits and our presence at the fair is an annual report to the people on how we're doing."

Missouri Department of Agriculture

In 1865, Missouri created a State Board of Agriculture with a two-fold purpose. First, it supplemented the work of local agricultural societies and product-based groups in providing information for Missouri farmers through the development of Farmers' Institutes. Second, it worked to encourage those amenities such as roads, bridges, and railroads that would make the farmer's occupation easier and more profitable. Norman J. Colman, a member of the State Board of Agriculture from 1865 through 1912, was a member of the first State Fair Board of Directors. The original state fair executive committee and officers were members of the State Board of Agriculture. During the early years of the twentieth century, the State Board of Agriculture and the University of Missouri cooperated to host the State Fair Boys' Schools as a means of teaching more modern agricultural techniques to young men.

In 1933, a reorganization of the state established the State Department of Agriculture. The Department's duties expanded considerably with the reorganization. In addition to encouraging environmental awareness and new uses for Missouri's agricultural goods, it regulates agricultural marketing and consumer protection. As agriculture has changed, the department, which from its beginning included horticulture and floriculture, has changed and now includes aquaculture.

The State Fair, whether under the direction of the Executive Committee, the State Fair Board, or the State Fair Commission, has always worked with the Missouri Department of Agriculture. In effect, the Missouri State Fair is a gigantic display of the Department of Agriculture's influence in all aspects of Missouri life.

Lieutenant Governor Edward Long and Commissioner of Agriculture John Sam Williamson enjoy a cartoon. (M. S. A.)

Below: The Fish and Game Department urged visitors to buy licenses at the 1927 state fair. (M. C. C.)

Missouri Department of Conservation

The Missouri Game and Fish Department, established in 1909, came as the result of over-harvesting of game and fish that had seriously depleted the state's wildlife population. The department mandated the purchase of hunting and fishing licenses, and used the resulting monies to raise and restock wildlife and maintain wildlife refuge areas.

Following the United States' entry into World War I in 1917, the Fish and Game Commission displayed wild fish and a specially designed "fish car" used to carry live fish to farm ponds for stocking at the state fair. The exhibit encouraged farmers to stock ponds and raise fish for their own consumption and for sale. The fish would provide a high quality protein without needing to be fed, thus freeing grain used for livestock feed to be sent to the front for the soldiers. In 1923, State Commissioner Frank Middleton supervised a state fair display of sunfish, bass, ring perch, catfish, shiners, silver perch, crappie, calico gold fish, carp, and buffalo in large glass aquariums. Ernest Schwarz of St. Louis presented an exhibit of stuffed birds displayed in front of paintings of their native habitats. In 1925, the department used a portion of the old swine building for its exhibit, which included two small deer.

In 1926, Commissioner Keith McCanse designed a pavilion of oak timbers for displays at the State Fair. The exhibit proved popular, and in 1928, added fish housed in separate aquaria by species. A large exhibit of hawks, including all species native to Missouri, supplemented the exhibit of indigenous birds and animals. The St. Louis Zoo brought live animals to supplement the exhibit of mounted animals owned by the department. The exhibit's supervisor praised it as "the most interesting exhibit at the fair," claiming there was not one "exhibit on the grounds which draws larger crowds."

The pavilion was repaired in 1933. Highlights of that year's exhibit included an eleven pound trout encased in a six-hundred pound block of ice housed in a special refrigerator, a duck hunting simulator that allowed visitors to shoot at images of flying birds, and a brooder house with one hundred quail chicks.

In 1935, the Works Progress Administration built a new wildlife pavilion from pink Missouri granite. 1936 marked the dedication by Commissioner Wilbur C. Burford and other officials of the new building with offices, lounging rooms, and storage space. A large aquarium in a setting of natural limestone displayed all species of Missouri fish. Baby animals, including timber wolf cubs, gray fox kits, half-grown wild

A young boy apprehensively pets a coyote. (M. S. A.)

Wild Animals Not Really Wild

When the State Fish and Game Department began showing Missouri wildlife at the fair in the 1920s, the animals were truly wild—orphaned or injured animals brought to the Fish and Game Commission offices, raised in captivity, and slated for release in the forests and prairies of Missouri. As Conservation agents learned more about the habits of wild animals, they realized that animals raised in captivity were poorly equipped to survive in the wild. Injured animals were often unable to survive in the wild even after their injuries had been treated. The animal exhibit at the Conservation Pavilion now reflects the best current information about animal survival.

Injured animals and birds reported to the Conservation Department are treated by veterinarians and naturalists at rehabilitation centers licensed by the state. After they are healed, the animals are taken to nature centers where their well-being can be assured and where they may be viewed.

During the state fair, selected animals are moved from the nature centers to the Conservation Pavilion, where they receive the best of care. Birds and reptiles are housed in a climate-controlled environment. All animal cages are cleaned daily, and the animals are provided with fresh water and appropriate food. After the fair, the animals are returned to the nature centers.

The casting pool was a great place to cool hot, tired feet. (M. S. A.)

turkeys, and albino skunk kittens, drew many visitors, as did white and gray squirrels, pheasants, turtle doves, sod hill cranes, wild ducks, quail, deer, and a raccoon who liked to place ice on his abdomen after eating. Old Blue, a large catfish, became a popular feature of the wildlife exhibit.

This building, repaired in 1952, expanded in 1959 to include a casting pool as part of an emphasis on sport fishing. The casting pool became a popular place for anglers learning casting techniques. Casting contests added to the interest in fishing. The building was expanded in 1986 to include a theater and a fifteen thousand gallon aquarium and two additional fish tanks. The Conservation Pavilion was further renovated with native landscape plants in 1995–97. It remains one of the most popular buildings on the grounds, drawing 70,000 visitors in 1945 and over 100,000 visitors in 2002.

Visitors enjoy watching Missouri fish in the aquarium. (M. C. C.)

Left: Woody Bledsoe and the Ozark Smoke Eaters entertain at the Conservation Pavilion in 1953. (M. C. C.)

The building's exhibits reflect changing attitudes and contemporary concerns of the Conservation Commission. In 1942 during World War II, the Game and Fish Department made displays emphasizing the importance of Missouri forests to the war effort. Fire prevention became a major focus of the building's displays. During the 1950s, fire prevention again became a focus as Missouri suffered a serious drought and the threat of forest fires was a major concern. For a number of years, Woody Bledsoe and the Ozark Smoke Eaters sang while Smokey the Bear cautioned people to be wary of starting forest fires.

In the 1960s, with the Conservation Department's push for farmers to develop wildlife habitats, the exhibits came to focus on wildlife, especially fawns, turkey, fish, and snakes. The Conservation Sales Tax passed by the voters in 1976 enabled the department to increase its efforts on behalf of conservation and ecology education. The department's exhibits at the fair continue to educate Missourians about the outdoors. In 1995, hands-on exhibits that allow children to pet animals and to watch working artists opened in the auditorium. Conservation

Department personnel visit with fair visitors, discussing wildlife and wildlife regulations. Smokey the Bear still talks to children, warning them that "only you can prevent forest fires."

Missouri Department of Natural Resources

The Missouri Department of Natural Resources was established in 1974 to "serve all the state's citizens through its involvement in resource and environmental issues." The DNR works to protect air, land, and water resources, to preserve historic resources, and to develop mineral and energy resources safely.

The Department of Natural Resources began demonstrating its projects in the Agriculture Building at the fair and in the Conservation Building. In the 1980s, the DNR acquired the Womans Building on the fairgrounds for its use. Archivists from the Missouri State Archives and volunteers from Sedalia boxed and sorted the mass of records that had been stored in the basement. The records were sent to the State Archives in Jefferson City where they were preserved and catalogued. The State Historic Preservation Office is supervising the ongoing restoration and renovation of the building.

DNR maintains exhibits showcasing its work in water quality, recycling, environmental protection, sustainable agriculture, and air quality. The exhibits draw visitors in to participate as well as watch. Tractors using soy-based fuel and solar cars appear on the lawn. Visitors collect items such as rulers and pencils made from recycled materials. Puppets and clowns encourage children to recycle and conserve. Children crawl through a tunnel that simulates the earthworm's underground tunneling. The DNR's environmental emergency response boat and vehicle demonstrate how they respond to spills of hazardous substances.

DNR also reminds visitors of the importance of caring for Missouri's past. Music groups perform traditional music—ragtime, bluegrass, and fiddling—on the porch as listeners gather on the lawn. The Missouri Archaeological Society teaches about Missouri's prehistoric past with its displays of artifacts and fossils. The Historic Preservation Office showcases their works by displaying photos

Rangers from Missouri's State Parks discuss their work with visitors. (V. B.)

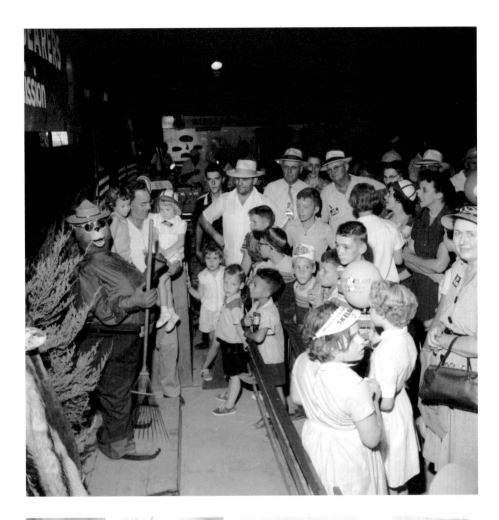

Smokey the Bear urges kids to help prevent forest fires. (M. C. C.)

Two young boys watch the display of mining techniques at the Department of Natural Resources display. (V. B. and M. D. N. R.)

A very tall Uncle Sam talks to children in the Womans Building. (V. B.)

of historic buildings that have been rescued from the wrecker's ball, renovated, and are being used as comfortable homes, community centers, and business buildings.

The Womans Building now houses the newly created State Fair Museum on the ground floor. The museum's theme, "Farming, Family, and Fun: One Hundred Years at the Missouri State Fair," highlights the history of the fair. Thompson Museum Associates created plans for exhibits presenting messages about the fair as a "showcase of rural traditions and values" and about the fair's mission to "celebrate and promote Missouri farming." Other exhibits highlight the growth of the grounds and buildings, and the extent to which fair "fun comes in all forms"—food, rides, Missouri products, sideshows, professional entertainment, and souvenirs. The Museum, dedicated during the fair in August 2002 by Governor Bob Holden, attracted thousands of visitors its first year.

University of Missouri and other State Colleges

Early fairs often awarded prizes to the largest or most unusual product, favoring what fair historian Wayne Campbell Neely calls the "freakish and useless." Neely considers agricultural colleges "a potent force in recent revisions of fair classifications" away from the unusual and toward the best examples of a standard product. As markets created a demand for agricultural products, agricultural colleges and experiment stations attempted to produce what the market demanded. These market demands in turn created categories of exhibits at fairs. Neely sees the cycle as complete: "The modern day fair works as interpreter of standards," offering examples of perfection to which all can aspire. Since the fair's inception, the University of Missouri's College of Agriculture has been involved in the Missouri State Fair. However, before the University could begin exhibiting at the fair, it first had to develop its agriculture programs.

The process of developing agriculture education in Missouri was long and difficult. Although science professors had attempted since 1858 to create agriculture classes at the University of Missouri, the College of Agriculture, desig-

Top: A crowd congregates on the lawn of the Womans Building to listen to old-time music. (V. B.)

Equipment is in place for the renovation of the Womans Building for the State Fair Museum. (V. B.)

In the 1950s, Joe Stiles of Collins, Missouri, found an excellent way to judge watermelons. (M. S. A.)

nated for "the branches of learning as are related to agriculture and the mechanical arts," was not added to the university until 1870. The newly developed college was hampered by the lack of learning materials; the few existing textbooks on agriculture were more related to climate and soil conditions in the east, and scientific experiments to determine the best methods of crop production had not yet been done extensively in the Midwest. In 1877, the Hatch Act, sponsored by Missouri Congressman William Henry Hatch and encouraged by then U.S. Commissioner of Agriculture Norman J. Colman, authorized and funded the creation of agriculture experiment stations, which enabled the College of Agriculture to test the best agricultural methods.

The university programs developed slowly, however, for the legislature was unwilling to provide adequate financial support and farmers were hesitant to send their sons to college and reluctant to change their established methods of production. Between 1895 and 1900, only an average of two students graduated per year. In 1901, the first year of the Missouri State Fair, enrollment in the College of Agriculture was approximately 125.

The university saw the state fair as a way to reach more farmers with the message of scientific agriculture. It designed its exhibits to educate farmers about new methods of planting, fertilizing, and harvesting. In 1928, president of the Missouri State Fair Board W. E. Leach called the fair "a condensed school of agriculture . . . conducted one week each year . . . Every farmer—no matter how young or how old—is afforded opportunities to learn some new farm methods, something that may possibly mean many dollars saved, new dollars for the farmer's treasury and a mitigation of back-breaking labor."

Professor Thomas I. Mairs, one of three crop sciences faculty members, demonstrated the effectiveness of cowpeas and soja beans as forage crops. The soja bean was a new crop; the 1904 *Encyclopedia Americana*, reported that "only within recent years"

"The Loveliest Party Imaginable"

Sedalia's Garden Clubs have long been involved with the Missouri State Fair. Members compete for prizes in the Floriculture exhibits and demonstrate new techniques in growing, grooming, and arranging flowers. For many years volunteers from the garden clubs maintained the flowerbeds in the boulevards leading toward the entrance to the fairgrounds.

The beautifully landscaped area of the Highway Gardens seemed to be a perfect place to recognize Missouri's Garden Clubs. The "Society and Clubs" page of the *Sedalia Democrat* described the 1935 garden party in the effusively feminine prose typical of journalism of the time, a time when married women, no matter what their own accomplishments, were identified by their husbands' names.

Mrs. Frank Leach, Mrs. R. R. Highleyman, and Mrs. J. E. Mitchell, with the four presidents and twenty-three representatives from the four Sedalia Garden Clubs "enthusiastically" made "elaborate plans" for what the society reporter predicted would be "the loveliest party imaginable." The local clubs planned to honor Missouri's First Lady Mrs. Guy B. Park; Mrs. S. F. Freeman, Springfield, state president of the Federation of Garden Clubs; Mrs. F. W. Sayers of Jefferson City, wife of Mr. Sayers of the maintenance department of the fair; Mrs. Charles W. Green, wife of the secretary of the fair; Mrs. G. Moore Greer, Sikeston, state fair hostess; and presidents of garden clubs in Missouri.

The ladies divided the Highway Gardens into four sections and assigned a garden club to take charge of decorations and refreshment tables in that section. Each club chose a color scheme and coordinating flower; Club1 selected brown and yellow with black-eyed Susans; Club 2 used lavender and yellow with jonquils; Club 3 chose pink, green, and white with sweet peas; and the newly organized Club 4 used blue, pink and white with pink rosebuds. The society reporter, who happened to be a member of Club 3, assured readers that its section would "doubtless be most artistic." Miss Hazel Lang and Mrs. Clarence Carter made nametags in the shape of flowers. A "bevy of pretty girls in colorful frocks" served fruit punch.

The guests of honor and chairmen of committees and club presidents formed a receiving line where they greeted visitors. The invitation was extended so "everyone interested in flowers and gardens is most cordially invited to enjoy the hospitality of the occasion." Visitors who attended the "delightful informal program" heard "notable speakers," listened to singing, and saw dancers from the Lylian Hurley Dancing School.

University of Missouri exhibited prize-winning sheep. (M. S. A.)

had its "value as a feed for stock been brought to the attention of the western countries." The cowpeas experiments would continue through the first decade of the twentieth century at the University. The soja bean, now called the soybean, would eventually become a major crop for Missouri farmers. By 1913, the University's exhibits had been recognized by the fair's Board of Directors as "a place to secure definite scientific information on farming."

In 1905, the Poultry Building was converted into a Dairy Building, and in the 1920s was renamed the University Building. In 1933, the University mounted an exhibit entitled Agriculture Economics with fourteen large booths. It included exhibits of lespedeza and legumes, mounted specimens of insects, an antique plow, a study of leg weakness in chicks due to deficient rations, demonstrations of methods of applying limestone and methods of candling eggs. The University also advertised its general university program, its college of engineering, the journalism school, and the crippled children's correction division.

The Missouri State Veterinary Association, using a building near the University Building, highlighted the work of veterinarians and encouraged the practice of veterinary medicine. In 1937, the Veterinary Building housed a theatre that showed films on agricultural subjects. Dr. M. E. Gouge and Dr. George Townsend, both of Sedalia, were in charge.

The College of Agriculture and the School of Veterinary Medicine of the University of Missouri now exhibit in the Agriculture Building, both demonstrating advances in their sciences and encouraging interested students to attend.

In 1870, the University of Missouri opened the School of Mines and Metallurgy in Rolla. This was the first technical institution west of the Mississippi River. This land grant school attempted to meet the "acute need for scientific and practical education." In 1915, the School of Mines hosted an exhibit highlighting the career possibilities in mining engineering featuring professors H. M. Thornberry, D. H. Radcliffe, and J. J. Dowd.

Other state colleges also exhibited at the state fair. Lincoln Institute, established in 1866 using donations from African American Civil War veterans, became a state-funded institution in 1879. In 1915, Lincoln won recognition for joinery and cabinet making, woodturning, working drawings (drafting), forge work, hand sewing, and wearing apparel for women. The 1957 Entry Book shows paintings by Lincoln University art students entered in the Fine Arts Department. Lincoln's exhibits at the state fair were particularly important because the racial segregation mandated by both law and custom in the first half of the century made it difficult for blacks and whites to see and learn from one another's accomplishments.

Kirksville Normal School, now Truman State University, was established in 1867 as the first of a number of regional teacher-training institutes. Warrensburg Normal School, now Central Missouri State University, was the second of the regional institutes, dating to 1871. These schools soon expanded to include programs other than teacher education. Both Kirksville

A solar car made by students at the University of Missouri-Rolla demonstrates the technology of the future. (V. B.)

The Rural Electric
Cooperatives exhibited in the
former Veterinary Building.
(M. S. A.)

Below: The Missouri State
Nurses' Association high-
lighted its work with a display
in the Education Building.
(M. S. A.)

and Warrensburg schools exhibited student work at the fair. In 1912, Warrensburg Normal School hosted cooking demonstrations in the Womans Building basement and "a splendid exhibit" in the Educational Department. The 1914 Entry Book shows twenty pieces of sewing and wearing apparel entered by Warrensburg Normal School, and in 1915, Warrensburg won for complete sets of three books in bookkeeping, art metal work, wearing apparel for women, and forge work. The Board of Directors praised the 1912 exhibit by Kirksville Normal as "constantly crowded with visitors" and the cited 1914 "lectures and demonstrations" given by Kirksville Normal for helping "to make the Missouri State Fair a permanent factor in up-to-the-minute, modern education." Now the state universities display bright, attractive exhibits advertising their campuses, their academic programs, student life, and athletics.

Missouri State Institutions

Missouri institutions for young people provided two types of exhibits at the state fair. First, they presented exhibits about their work with youth. Second, they displayed products made by their inmates. The Industrial Home for Girls at Chillicothe had been established in 1887 as a reformatory for delinquent girls, though many of the girls sent there were orphaned or neglected, rather than actually delinquent. It described its purpose, according to historians Lawrence Christensen and Gary Kremer, as providing a place where girls might be "removed from vicious associates and evil influences," and be "reformed and become good domestic women, prudent in speech and conduct, cleanly, industrious and capable housekeepers." As a result, half an inmate's day was spent learning practical skills such as housekeeping, laundry, gardening, and sewing.

Fine hand sewing was particularly prized, and inmates entered their work in competition at Missouri State Fairs. In 1923, for example, girls from the Industrial Home exhibited and sold needlework items. In 1933, the home again exhibited needlework and basketry. The *Sedalia Democrat* praised the exhibit, saying, "The display is interesting, not only because of the beauty and usefulness of the articles shown, but it is interesting to know that girls from all parts of the state, sent to this school for correction, are taught to do this sort of work."

Visitors admire hand sewing and home baked goods in the Home Economics Building. (M. S. A.)

The General Assembly also authorized in 1887 the establishment of the Reform School for Boys, later called the Training School for Boys, at Boonville. While some of the boys had been convicted of crimes, many of them, like the inmates of the Industrial Home for Girls, were unwanted or neglected. The Training School was self-supporting, and *Somewhere in Time: A 160 Year History of Missouri Corrections* reports that the boys did "all the cooking, baking, tailoring, and gardening." Boys were taught trades such as shoemaking, blacksmithing, electrical engineering, brick making, and farming. In 1915, the school won awards for basket making, colored maps, sewing six garments, forge work, and a map or chart used in teaching.

Missouri state institutions displayed goods made by their residents in the Varied Industries Building. In 1937, when Missouri was still suffering the effects of the Great Depression, questions were raised as to whether the expense of shipping such exhibits was justified. A decision to ask for exhibits came a week before the beginning of the fair. State institutions for the handicapped, including the State Hospital at Fulton, the State Hospital at Nevada, and the State Hospital for the Mentally Handicapped at Marshall sent exhibits. The Entry Book of 1957 shows that the State Hospital at Nevada showed pillowcases, handbags, crocheted baby clothes, place mats, and dolls made by residents. Since the mentally handicapped were often kept isolated at this time, exhibits at the fair provided visitors a rare opportunity to see their work.

Missouri Highway Patrol

In the early years of the Missouri State Fair, the fair secretary named a fair police chief who was assisted by representatives from police forces throughout the state. Sedalia police officers provided crowd control outside the grounds, and at the first fair, closed a concession stand serving beer located right outside the fairgrounds. When extra security was needed, as it was when President Taft visited, extra officers were requested from metropolitan police forces.

In 1931, Governor Harry Caulfield signed Senate Bill 36, which established the Missouri Highway Patrol as a state police force, and named Lewis Ellis as its superintendent. The Patrol purchased several vehicles, including thirty-six Model A Ford Roadsters costing $413.18 each. Equipped with Klaxon horns, spotlights, and lighted "Patrol" signs in the windshields, they provided the new patrolmen a means of apprehending lawbreakers and monitoring traffic regulations. Their fire extinguishers and first aid kits enabled the patrolmen to assist in emergencies. Troopers drove with the tops down so other drivers would be aware of their presence. As the cars had no heat, patrolmen were uncomfortable as well as dedicated.

In 1933, for the first time, twelve patrol officers served at the Missouri State Fair. As the fair grew, the number of patrolmen assigned increased. In 1937, they also supervised traffic on the highways leading into Sedalia. The Highway

Police officers from throughout the state pose in front of the Administration Building. (S. D.)

Below: Highway Patrol officers and two Kansas City Police Department officers began a bicycle patrol of the fair in 1995. (M. S. H. P.)

Otto the Talking Car chats with a young visitor as Corporal Paul Reinsch observes. (M. S. H. P.)

Patrol then maintained its headquarters at State Fair police headquarters just south of the grandstand. The State Highway Patrol continues to provide traffic control in Sedalia at fair time, police the grounds for lawbreakers, and provide security at concerts. In 1995, ten officers began a bicycle patrol at the fair.

In the 1960s, the Highway Patrol built a metal building on the southwest part of the fairgrounds. The building houses Otto the Talking Car, who first appeared at the fair in 1969. Otto is a 1936 Ford Roadster, one of the original patrol vehicles. He wears a patrol hat above his smiling fiberglass face. He greets visitors, especially children, blows his horn, blinks his eyes, wipes his windshield, and talks to children about safety.

Missouri Department of Transportation

As automobiles became more widely used and more affordable, Missouri became more interested in the quality of its roads and the Missouri Department of Transportation came into being. Under Governor Joseph Folk, the state began a program of road and bridge construction. The Executive Committee of the State Fair Board reaffirmed in 1906 that the "greatest value of the state fair is its educational features," and suggested the fair teach the benefits of good roads. In 1915, the Missouri State Fair celebrated Good Roads Day. Governor Elliot W. Major spoke on the importance of good roads and State Highway Engineer Frank W. Buffman gave pointers to a group of advocates of good roads. A parade of road-making machinery, including scarifiers, road graders, and planers supervised by County Highway Engineer T. A. Stanley, demonstrated how to build good roads. The demonstrations were filmed to be shown at theaters throughout the state. In 1918, the State Highway Commission demonstrated tests of road materials, methods of road construction, and explained its work in an exhibit in the Swine Building.

In the 1920s, the Highway

Right: The beautifully landscaped Highway Gardens provide a natural setting in which to relax. (M. S. A.)

Department built the Highway Gardens as a tribute to the roadside parks that dotted the Missouri Highway System. The Gardens feature a pond with water lilies and goldfish, water fountains set into a rock, benches and picnic tables, and plantings. They remain a favorite place for visitors to rest in the shade, have a drink of ice water, and enjoy the beautiful landscape. In the spring and summer months, they are a popular site for outdoor weddings. In the 1930s the Sedalia Garden Clubs hosted garden parties at the Highway Gardens and invited state dignitaries for punch and cookies.

In 1935, the Highway Gardens' indoor exhibit focused attention of highway safety, and included twenty-four exhibits on such diverse subjects as speeding, carbon monoxide, train accidents, children's safety, and "road hogs." The color scheme used in the building and fixtures was based, according to Assistant Maintenance Engineer F. W. Sayers, on a scheme using loud speakers and neon lighting used at the Chicago World's Fair in 1933–34. Later, the Highway Gardens' indoor exhibit became a theater showing films about Missouri.

The Missouri State Fair is indeed a showcase of Missouri. As the various state agencies demonstrate their work, they remind visitors of the vast array of services offered by the state and of the immense talent that lies within Missouri.

The Grand Opera House at the Highway Gardens showed movies of Missouri scenes. (M. S. A.)

CHAPTER 5

"WHAT IT IS POSSIBLE TO ACCOMPLISH":

Showcasing Youth Work at the Fair

A girl plays with a day-old kid at the fair. (M. S. A.)

THE TURN OF THE CENTURY MARKED A CHANGE IN demographics in Missouri and a change in attitude toward education, both of which affected the Missouri State Fair. At the same time agriculture was becoming more mechanized and scientific, youth were leaving the farm because of the difficult lives they saw there and the lures of city life. The percentage of Missourians living on farms dropped from 63 percent to 53 percent during the first two decades of the twentieth century. Education changed during the Progressive Era, as the first two decades of the twentieth century are known, in part to counteract the flight from the farm.

Influenced by the ideas of John Dewey—that children learned best through doing, that they learned best when new material was presented in relation to something they already knew, and that they learned best when material was presented in a manner geared to their intellectual development—schools slowly shifted away from pure "book learning" to more relevant information and skills. Under pressure to teach the "whole child," city schools added industrial arts and commercial classes to teach boys and girls workplace skills.

Rural schools adapted more quickly to the idea of practical education, as the National Patrons of Husbandry—the Grange— had been campaigning since 1874 to add practi-

This exhibit explains the benefits of Vocational Home Economics. (M. S. A.)

cal agriculture and domestic science to the rural schools. As a result, in the early twentieth century, rural children were learning those things that would benefit them in their future lives as farmers and farm wives. While the concept of vocational education was often criticized as trapping children in certain roles, it did fulfill one of its goals: by making farm work easier and more profitable and by making farm life more attractive, farm youth were more inclined to remain on the farm.

Nature study provided an introduction to the sciences in an easy and practical manner. Instead of simply reading about plants and animals, students would plant seeds, gather and identify leaves, observe birds, and monitor the

Greg Grupe shows off his
insect collection. (M. S. A.)

The Callaway County Calf
Club shows its calves.
(M. S. A.)

The Boone County Calf Club exhibits its stock. (M. S. A.)

life cycles of insects. Mrs. M. T. Harvey of Northeast Missouri State Teachers College, now Truman State University, introduced the concept of nature study to Missouri schools when in 1900, she included a nature study component in the teacher education curriculum. Soon nature study moved into agriculture study, based on the idea that crop yields could be improved by applying scientific principles of testing various methods of planting, fertilizing, and cultivation, recording the results, and then applying the best methods to future crops. State School Supervisor R. H. Emberson encouraged the teaching of agriculture in rural schools, and in 1910, the Missouri legislature established agriculture classes as a part of the curriculum in Missouri high schools. The Smith-Hughes Act of 1917 gave federal money to rural schools for vocational education. Two youth programs—4-H Clubs and the Future Farmers of America—grew out of the introduction of agriculture and domestic science classes to the schools.

Boys and girls agriculture clubs developed as teachers across the country extended instruction outside school hours. Schools and civic groups sponsored these clubs, which took the names of projects; "corn clubs" for boys and "tomato clubs" for girls became common. A typical corn club project involved a boy's taking charge of an acre of land and growing a corn crop using the best methods. Tomato clubs involved a girl's adopting a portion of an acre of tomatoes and raising a crop, then preserving the produce. "Pig clubs" and "calf clubs" introduced the latest livestock management techniques.

School Exhibits

Many schools held fairs in which their students demonstrated the practical things they produced. For example, in 1911, High Point School in rural Pettis County held its fourth annual fair in which students displayed breads, doll clothes, quilts, vegetables, and field crops. The school's exhibit was praised as an event that could show students "what it is possible for them to accomplish."

Schools also exhibited their students' work at the Missouri State Fair. Individual schools made exhibits of work done by their students during the previous school year. These exhibits were displayed in various buildings, including the Art Building and the Education Building. The 1914 exhibit, according to the Secretary's Report, "showed the marked progress of the

The Howard County Calf Club included a young woman as one of its members. (M. S. A.)

Arator School in rural Pettis County displayed its students' work in 1926. (P. C. H. S.)

Lincoln-Hubbard School in Sedalia provided elementary education for African American students in Sedalia and high school education for a five-county region; it exhibited its students' work at the fair. (M. S. A.)

schools and educational institutions in the state." In 1915, the *Sedalia Democrat* praised the exhibits of crayon drawings, writing books, pencil drawings, watercolor paintings, agricultural notebooks, collections of insects, and a mounted collection of useful seeds. In 1923, school exhibits included a model

Missouri Synod Lutheran Schools advertise their schools. (M. S. A.)

dollhouse, hand sewing, a kitchen cabinet, posters, costume designs representing clothing from many nations, and artwork borders for holidays. In addition, individual counties presented demonstrations or programs of readings and music by students selected from their rural schools, such as the program presented in 1937 by the county schools of Pettis County. Students from Tanglenook School, Georgetown School, Camp Branch School, Arator School, Bothwell School, and other rural schools sang, played rhythm instruments in a rhythm band, played piano solos and duets, and gave readings.

Scouting

Boy Scouts and Girl Scouts offered young people opportunities for nature study, camping, outdoor activities, and community service. In these organizations, boys and girls could master citizenship and leadership skills and demonstrate their competence by earning merit badges in a variety of fields. Established in Great Britain in the late nineteenth century, Boy Scouting came to the United States in 1910. During World War I, Boy Scouts assisted in the war effort, and participated in the Missouri State Fair with the county Councils of Defense, the Red Cross, the Knights of Columbus, and "all kindred war work bodies" in a "big patriotic parade" held on Sedalia Day. In 1923, 120 scouts from the Boy Scout Band in Springfield appeared as one of that year's "big amusement features." The band's repertoire included classical numbers such as the "William Tell Overture" and "Tannhauser" as well

Boy Scouts and Girl Scouts used this building in the late 1940s and early 1950s. (M. S. A.)

as popular songs.

In 1945, Boy Scouts became a separate department at the Missouri State Fair. Classes of exhibits included troop projects such as scrapbooks and individual projects such as woodworking, bookbinding, leather craft, first aid, and Indian costumes. Sea Scouts demonstrated knots and Air Scouts showed homemade sextants. The Boy Scout Department at the state fair ended in 1951, but scouts continued to be recognized on Boy Scout day, when members wore their uniforms to the fair.

Girl Scouts were started in Georgia in 1912, and quickly spread across the nation. Girl Scouts participated in early fairs; in 1925, they exhibited at the Education Building, showing posters about their different activities. The *Sedalia Democrat* noted that the exhibit was

Boy Scouts enjoy a drink at the Dairy Bar with Lieutenant Governor Long. (M. S. A.)

"attracting much attention." In the 1930s, Girl Scouts rented exhibition space and demonstrated handicrafts, childcare, laundering, and first aid. They also sold cookies.

Girl Scouts became a state fair department in 1942. Classes included scrapbooks documenting troop community service projects, puppets, clubroom decoration, and grew to include clothing construction, homemaking, arts, crafts, and baking. Girl Scouts remained a fair department until 1953, but also continue to participate in the fair by attending in uniform on Girl Scout Day.

State Fair Boys' School

In 1907, the Missouri State Board of Agriculture sponsored boys' schools. Led by Sam Jordan, the "Missouri corn man," these schools brought boys from throughout the state to learn the best methods of increasing corn crops. Though the first boys' schools were held at private farms, in 1912, they moved to the Missouri State Fair. In 1914, over one hundred boys attended; in 1915, the number had risen to two hundred. Boys between twelve and seventeen competed in a test of arithmetic, geography, and agriculture, and in an essay contest administered by the various county school superintendents to earn a scholarship that paid their expenses while attending the Missouri State Fair Boys' School; one boy from each of Missouri's 114 counties won the privilege of attending at state expense. The 250-word essay reflected agricultural issues of the day; the topic in 1918 when the United States was involved in World War I, for example, was titled "The Farm Boy's Part in Winning the War." The boys' families were encouraged to attend the fair while their sons attended the school, but parents who could not attend were assured by the *Ruralist* that "the small chaps are cared for in an admirable manner."

In 1923, State Superintendent of Schools W. G. Dillon and state high school Inspector R. K. Phelps supervised the school. The boys camped in tents on the south end of the fairgrounds and met each day for demonstrations and instruction in all aspects of agriculture by nationally recognized experts. In 1925, again under Dillon's direction, the school provided lectures by faculty from the University of Missouri and members of the Missouri State Board of Agriculture.

A boy at the State Fair Boys' School learns about livestock judging. (M. S. A.)

Jersey cattle raised by young people are judged in the Coliseum. (M. S. A.)

This calf club specialized in Hereford cattle. (M. S. A.)

In 1933, the economic problems of the Great Depression limited the number of attendees to eighty-five, and in 1934 to seventy-five. By 1935, attendance had increased to 102. Evert Keith, superintendent of that year's school, praised the school for its "broadening of experience and social development, the big place where rural boys have been neglected in the past."

Fair historian Wayne Campbell Neely comments, "The educational function performed by the competitive display of agricultural products in creating standards and interpreting them to the public is revealed at two distinct points: the classification of products and the judging of exhibits." One of the goals of the boys' and girls' clubs was learning the criteria by which top quality products—livestock, crops, vegetables and fruits, clothing, baked goods—should be evaluated. The Missouri State Fair Boys' School encouraged careful evaluation of farm products as well; in 1915, the Boys' School awarded prizes to the best judges of corn. In 1918, livestock judging was one of the featured activities. Another goal of both the boys' and girls' clubs and the Boys' School was the production of quality products. For example, in 1906, the fair held a Boy's Corn Contest. Yet another goal was developing the ability to demonstrate proper production techniques to others. In addition, these clubs and the Boys' School provided a way of reaching parents with new ideas by having young people demonstrate that new methods could be successful.

4-H at the Fair

The boys' and girls' agriculture clubs became the basis for the 4-H Clubs. By 1911, the four-leaf clover had been adopted as the 4-H emblem, though many clubs were still called boys' and girl's clubs. In 1914, the Smith-Lever Act passed Congress and created the Cooperative Agriculture Extension Service, whereby federal, state, and county governments worked together. The act provided a way for the Morrill Act colleges to "extend" university programs to all people. County extension agents, trained in agriculture and domestic arts, advised farm men and women about new techniques and also assisted boys' and girls' club leaders. According to historians Lawrence Christensen and Gary Kremer, by 1916, over eleven thousand Missouri young people were in agriculture clubs. In 1927, 4-H Clubs became part of the extension program.

The Missouri State Fair first offered prizes for boys' and girls' club work in 1918, awarding a total of $335.55 in prize money. Demonstrations of canning focused on preserving food for use at home as part of the war effort; that year's slogan was, "We pickle and can for Uncle Sam." Other activities included breads, corn, poultry, calves, sheep, pigs, garden exhibits, draft

Missouri Ruralist recognized Mr. and Mrs. Clifford Diamond of Harrisonville for their combined thirty-one years of work as 4-H leaders. (M. S. A.)

horses, beef cattle, dairy cattle, and swine. Boys' and Girls' Club exhibits continued to expand; in 1921, 150 members from twenty-five counties exhibited. Neely notes that agriculture magazines recognized the relationship of club work to fairs; in 1920 the *Ohio Farmer* commented that, "the boys and girls' club work is the life blood of the present day fair."

Boys' and girls' club work at the fair increased as home management categories were added in 1922, poultry judging in 1923, a health contest in 1924,

and dress revue in 1925. Prize money awarded also increased, from $2,617.75 in 1922 to $3,763.34 in 1924. In 1925, the name 4-H was added to the fair's identification of the Boys' and Girls' Club Work division, and in 1928, the division became the Boys' and Girls' Club 4-H Department. In 1928, 4-H contests moved to the University of

Left: Governor Smith congratulates Gerald Fuchs of the Lafayette County 4-H. (M. S. A.)

Waltman Miller of Granger, Missouri, owned the first prize 4-H Shorthorn steer and the Grand Champion, Junior Division, Shorthorn steer. (M. S. A.)

Below: A mounted 4-H Club lines up for a parade. (M. S. A.)

Missouri and the state fair became the place for exhibiting the best of Missouri 4-H Club work.

Agricultural journals and historians recognized the importance of 4-H work to both fairs and agriculture. In 1931, C. E. Cameron, president of the *Iowa State Fair Board*, said "in the fostering of boys' and girls' club work alone, and in providing a medium of competition for these junior farmers, our fairs of the last decade have made a contribution to the future of agriculture, the magnitude of which no one now living can fully estimate." Neely points out that by 1935, "club activities at practically all fairs have become one of the leading parts of the program."

THICKSET
1st--4-H Shorth
Steer
Grand Champion,
Junior Divis
Owned by Waltma
Miller, Gran

In 1938, the work represented by the boys' and girls' clubs became the Junior Activities Department, which included work by both 4-H members and Future Farmers of America. Livestock shown in the Junior Activities Department competed for $4,586.50 in premiums. In 1952, 4-H Club work became recognized as the 4-H Department. 4-H work expanded to include

projects that would benefit the entire farm family—electricity, woodworking, home decoration, and landscaping. The number of exhibits and classes continued to increase.

In 1963, entries included 1,563 in clothing, 1,086 in foods, 780 in home management, 231 in home grounds, 96 in dairy cattle, 216 in beef cattle, 152 in swine, 159 in sheep, 116 in electricity, and 437 in woodwork. 4-H members competed for $10,820 in premiums. 4-H Club work has expanded beyond agriculture to include forestry, shooting sports, outdoor skill, photography, lifetime sports, aerospace studies, and global education. The 4-H Department grew from twenty categories in 1984 to thirty categories in 1990, when 3,973 exhibits represented the work of clubs in ninety-six counties. In 2001, 7,451 exhibits constituted 149 classes in forty-seven categories such as soy foods and soybeans, and specialty ham curing, as well as the livestock, woodworking, electricity, cooking, and sewing contests of previous years.

In the early years, young people's activities had no specific place on the grounds. After the new Swine and Sheep Pavilion was built in the 1920s, a forty-foot by twenty-four foot section of the original swine and sheep building was partitioned and whitewashed to provide space for 4-H demonstrations. In 1940, 4-H beef exhibits moved into the Donnel Building. In the 1940s and early 1950s, 4-H Clubs used space in the University Building. According to "The History of 4-H at the Missouri State Fair," the University Building served for a few years as the 4-H Building. In 1955, the *Sedalia Democrat* reported that, "the exhibits from the 4-H clubs over the state are so extensive that they cannot be displayed in such a manner that the high quality of the work may be observed." The larger building, originally built as a poultry building in 1905 and later designated the Missouri Building, became the 4-H Building in 1958. In 1959, the Rural Electric Cooperatives helped create the Electric Theater, which enabled 4-H members to give demonstrations using electricity.

Girls with their ribbons and garments check the Program in 1959. (M. S. A.)

4-H Club work remains an important component of the modern fair. The young people who participate learn valuable skills and gain poise, self-confidence, and leadership skills. Historian Donald Marti suggests that, "4-H and its various allies have recalled agricultural fairs to their first, educational purposes."

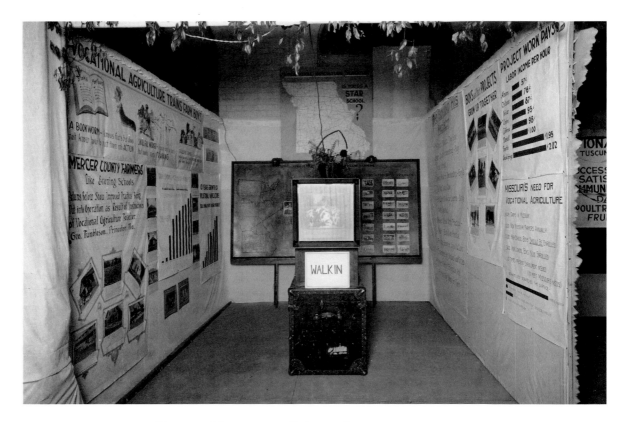

Vocational Agriculture became a part of Missouri schools in 1917. (M. S. A.)

A Vocational Agriculture student raised these prize-winning sheep. (M. S. A.)

Future Farmers of America

The Future Farmers of America also grew out of the vocational agriculture movement of the early twentieth century. The National Organization of Future Farmers of America originated in 1928 in Kansas City, Missouri, when thirty-three delegates from seven states met. By 1934, FFA had spread throughout the nation. In Missouri, FFA originally lay under the State Superintendent of Schools, with strong support from the College of Agriculture at the University of Missouri; it is now part of the Department of Elementary and Secondary Education.

The Missouri State Fair first recognized vocational agriculture students by creating a classification of exhibits under the direction of Superintendent W. T. Spanton in 1922. The *Sedalia Democrat* announced that $1,000 in prize money would be offered; students from seventy-six schools had promised enough exhibits "to make

a very creditable showing." Three times more work than anticipated arrived at the fair, making the exhibit in the Education Building "two hundred percent

Future Farmers of America Creed

I believe in the future of agriculture, with a faith born not of words, but of deeds—achievements won by the present and past generations of agriculturalists; in the promise of better days through better ways, even as the better things we now enjoy have come to us from the struggles of former years.

I believe that to live and work on a good farm, or to be engaged in other agricultural pursuits, is pleasant as well as challenging; for I know the joys and discomforts of agricultural life and hold an inborn fondness for those associations which, even in hours of discouragement, I cannot deny.

I believe in leadership from ourselves and respect from others. I believe in my own ability to work efficiently and think clearly, with such knowledge and skill as I can secure, and in the ability of progressive agriculturalists to serve our own and the public interest in producing and marketing the product of our toil.

I believe in less dependence on begging and more power of bargaining; in the life abundant and enough honest wealth to help make it so—for others, as well as myself; in less need for charity and more of it when needed; in being happy myself and playing square with those whose happiness depends upon me.

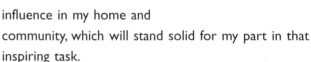

I believe that American agriculture can and will hold true to the best traditions of our national life and that I can exert an influence in my home and community, which will stand solid for my part in that inspiring task.

The Republic, Missouri, Future Farmers of America Chapter won the MFA Plaque for its beef cattle. (M. S. A.)

Future Farmers of America members display farm equipment they had made. (R. H.)

better than any ever held before." The largest exhibit came from Lincoln Institute in Jefferson City. Young men from vocational agriculture programs attracted large crowds by demonstrating soldering, corn judging, carpentry, rope making, spraying, and mixing spray, attracting large crowds. The Vocational Agriculture exhibits in 1927 were larger, featuring 161 hogs, seventeen baby beeves, sixteen sheep, and six dairy heifers. In addition to the animals shown by individual students, nineteen schools hosted displays. That year, Vernon Jelley, a sixteen-year-old Vocational Agriculture student from Santa Fe, Missouri, won first, second, and third prizes in the Junior Sow Pig Class, and first in the produce of dam, get of sire, futurity of litter and young herd. His Chester White sow was Junior Champion and three of her pigs were named best boar pigs in the Chester White show.

Vocational Agriculture remained a separate classification until 1931, when the *Missouri State Fair Premium List* identified a "Vocational Agriculture and Junior Project Work" class. In 1932, the *Premium List* first mentions Future Farmers of America, and fair regulations required that students exhibiting in the Vocational Agriculture classes be members of FFA.

The 1933 Vocational Agriculture exhibit was "larger than usual," said Superintendent R. T. Wright, with 250 hogs, 55 calves, and 65 sheep. The Sweet Springs School had the largest number of stock at the fair, followed by the Chariton County Schools of Keytesville, Salisbury, and Brunswick. Twenty vocational agriculture programs exhibited at the fair that year, with the Hatfield, Lexington, Liberal, and Norborne Schools showing for the first time. By 1964, the number of FFA members participating in the fair had increased; the *Missouri Ruralist* reported that over 550 members were on the grounds.

The major breeds of livestock and judging contests constituted the FFA activities during the early years of exhibition at the Missouri State Fair, but other aspects of agriculture soon came to be part of the FFA exhibits. Beef heifers were added in 1949, and the farm mechanics show in 1955. By 1961, the farm mechanics show included 284 projects. FFA began exhibiting field crops in 1964. In 1968, the field crop exhibits numbered sixty-eight; by 2001, the number of crop exhibits had increased to 1,891. During the 1990s, a Horticulture Class was added and in 1999, a Fruit and Vegetables class was added.

In 1937, Governor Phillip Donnelly created the Governor's Award to recognize the chapter with the best livestock display, based on an evaluation of the quality and attractiveness of the stock and the stables. Norborne FFA won the first Governor's Sweepstakes Trophy; it continued to win from 1938–1942, and again from 1948 through 1951. Some chapters such as those in Stet, Hamilton, Chillicothe, Aurora, and Bowling Green have repeatedly won this award, now called the Director's Sweepstakes.

The Missouri Farmers' Association began awarding plaques to outstanding chapters in 1975; the MFA Plaques identify individual classes of FFA work. Again, some chapters have dominated this award. The Troy Chapter won the Farm Mechanics MFA Plaque from 1975 through 1999; in 2000 and 2001, the Eldon Chapter won. Bowling Green has won the Swine MFA Plaque more frequently than any other chapter, winning in 1976, and from 1979 through 1999. The Hamilton Chapter has frequently won the MFA Sheep Plaque.

In 1955, the FFA moved into the former University Building, which had started its life as the Poultry Building in 1903 and became the Dairy Building in 1905, according to Ray Hagan, former state director of Future Farmers of

Chicks hatching in the incubator at the Children's Barnyard fascinate visitors. (M. S. A.)

This group of FFA members ushered in the Coliseum when President Reagan visited the fair in 1984. (R. H.)

America. The lawn to the west and south of the building held the machinery exhibits in which students showed the trailers, benches, and loading chutes they had made. In 2001, the 7,261 Future Farmers of America entries showed the public that the future of agriculture in Missouri is in capable hands. In 2002, a new FFA Building was opened to provide more space for the Future Farmers to show their work.

The Future Farmers of America serve in other ways at the Missouri State Fair. From 1960 until 1982, they ushered at the grandstand shows. In 1984, state FFA members ushered and assisted security personnel at the Coliseum when President Ronald Reagan visited the fair.

FFA is also responsible for one of the most popular exhibits on the fairgrounds—the Children's Barnyard, advertised in the *Premium List* as a "Children's Wonderland." In 1959, FFA chapters close to Sedalia created a display of adult and young livestock, including beef and dairy cattle, horses, mules, poultry, sheep, goats, swine, dogs, and cats. One of the most popular exhibits is the incubator containing hatching chicks; another popular exhibit is the huge boar. The Children's Barnyard provides many children from urban areas a chance to learn

4-H Pledge

I pledge my HEAD to clearer thinking,

my HEART to greater loyalty,

my HANDS to larger service

and my HEALTH to better living,

for my club, my community, my country

and my world.

about farm animals they might otherwise not get to see.

Both 4-H and FFA members participate in the Sale of Champions. The Grand Champion and Reserve Grand Champion animals produced by young people are auctioned to business leaders and agriculture companies who compete to buy the steers, barrows, lambs, poultry, and rabbits.

Future Farmers of America is essential to the continued success of American agriculture. As American farmers are being asked to produce more of the world's food supply and as family farms face increased competition from corporate farms, the young people of FFA learn the skills necessary to succeed. *Successful Farming* acknowledges the importance of FFA to state fairs, citing manager Craig Atkins of the South Dakota State Fair, "The FFA and 4-H programs are number one. They teach farm values, good sportsmanship, and responsibility."

Leroy VanDyke auctions prize-winning hams and animals. (M. S. A.)

CHAPTER 6

"POLITICIANS AND FAST HORSES":

The Missouri State Fair Through the Years

The "Show Me" Girl advertised the fair. (V. B.)

N OEL PERRIN WROTE IN THE *SMITHSONIAN* THAT EVERY fair had to have "politicians and fast horses." The Missouri State Fair has hosted many politicians—governors, senators, representatives, and mayors. The state fair has also played host to three presidents—Taft in 1911, Truman in 1955, and Reagan in 1984. The fair has also featured the "fastest horse in the world," races between an ostrich and a pacing horse, anniversary celebrations, the Wright Brothers, a number of balloons and dirigibles, parades, extravagant patriotic celebrations, and numbers of beautiful ladies. Journalists over the last one hundred years have described the events of the fair, *Program Books* have announced them, and participants have remembered them. Together, the highlights of the Missouri State Fair present a picture of changing times and changing attitudes.

A visitor is "flying high" at the 1912 fair. (V. B.)

Below: Arch Hoxsey flew the Wright Brothers' plane in a 1910 appearance. (V. B.)

Presidential Visits

The presidential visits, and information about them, reflect the issues of the day, the presidents' personalities, journalistic style, and concerns for security. When William Howard Taft visited, he justified his political stance and played golf; Reagan came, gave a campaign speech, and departed. Press coverage ranged from printing Taft's entire speech to briefly mentioning former president Truman's comments. Four secret service men accompanied Taft; when Ronald Reagan came, a corps of security personnel came, accompanied by a contingent from Whiteman Air Force Base, extra Highway Patrol officers, and local law enforcement personnel. Taft was accompanied by five press representatives from newspapers; Reagan attracted a press corps of at least seventy-five representatives from newspapers, radio, and television.

The Largest Meal and the Longest Speech: William Howard Taft

In the summer of 1911, President William Howard Taft embarked on a forty-six day tour of twenty-six states. In Missouri, he visited St. Louis and spent a day in Sedalia visiting the Missouri State Fair. The president's twelve-hour visit was well coordinated, according to the *Sedalia Democrat Sentinel*: "The various committees have left nothing undone to provide for his comfort and entertainment while here." In the typical effusive prose of the day, the Sedalia press described every aspect of the president's visit, from his progress through the fairgrounds to the decorations, the menu, and the charm of the hostesses at the Sedalia Country Club, where Taft was entertained.

The president arrived by train in his own Pullman Palace car, which parked on the tracks that ran onto the fairgrounds. The train was escorted into Sedalia and onto the fairgrounds by "daring aviator" Hugh A. Robinson, who flew his "Curtiss aeroplane at a furious rate of speed." Although its

Dirigibles frequently visited the fair; this airship arrived in 1908. (P. C. H. S.)

FIRE!

I n October 1903, most of the original frame buildings on the fairgrounds burned. Custodian Frank Beemer discovered a fire in Cattle Barn Number 1. Two Sedalia Fire Department hose companies responded, but the flames spread rapidly to a second cattle barn, then to the heavy horse barns, and finally to the fairgrounds fire department building, a public restroom, and thirteen M. K. and T. Railroad cars. The total loss of approximately $30,000 was not well covered by insurance, in part because the Board of Directors considered the insurance costs "exorbitant," as the rates they had been quoted were "five times what is charged on farm property" near the grounds, even though the fairgrounds had city water and was served by the city fire department. The buildings were replaced by brick and steel buildings.

technology had developed rapidly, the airplane was still a novelty. Governor Hadley ordered four mounted officers and two detectives each from Kansas City and St. Louis, and two detectives from St. Joseph to the grounds to insure extra security for the president.

The president was to arrive at 8:00 a.m. and have breakfast at the Sedalia Country Club, whose golf course bordered the fairgrounds on the north. Hostesses Mrs. John T. Stinson, wife of the fair secretary; Mrs. W. H. Reynolds; and Mrs. C. E. West had put "untiring efforts" into assuring that the president's entertainment was a "success."

A watch fob (front and back), announced Taft's visit in 1911. (V. B.)

Below: President Taft and his entourage drove onto the fairgrounds. (P. C. H. S.)

The president's visit began at breakfast at the country club at 8:45. Fair officials and the president's retinue, seated overlooking the golf course, dined on "Missouri style" fried chicken with crème gravy, French potatoes, cold tomatoes, hot biscuits with marmalade, broiled steak with mushrooms, and coffee, accompanied by cigars. The extravagant menu was typical of the time and a tribute to president Taft's appetite; at 325 pounds, he was the largest president. John Stinson's daughter Ruthie pinned a boutonnière on the president's lapel.

Taft had asked to meet and was introduced to Sarah Smith Cotton, daughter of Sedalia founder George R. Smith and an active supporter of the Republican Party. Taft toured downtown Sedalia at 10:00 a.m. and arrived back at the fairgrounds at 10:45. Hiner's Band of Kansas City, the 1911 official state fair band, played "The Star Spangled Banner" and 125 cadets from Kemper Military Academy cheered the president, accompanied by the Kemper Band and led by Cheerleader Frank Purvis. The president appeared at the Coliseum, where he was introduced by Governor Hadley, who had in turn been introduced by fair President W. A. Dahlmeyer.

The local press printed the text of Hadley's brief introductory welcome,

and the extremely long speech by the president. Taft praised the Missouri State Fair, emphasized that "the farmer is coming into his own" as agriculture became more scientific and profitable, and explained at length his recent political decisions, including calling an extra session of Congress and trying to influence the details of the controversial Payne-Aldrich Tariff legislation.

At 2 p.m. the president arrived at the grandstand where he watched a parade of prize livestock, special races, and an aeroplane exhibition. Smithton mule breeder Louis Monsees arranged a "Presidential Mule Show" of mules from throughout Missouri led by men in identical white suits and caps. After the stock show, the president and his party moved to the Sedalia Country Club, where he golfed with Governor Hadley. The president was, despite his bulk, thought to be "exceedingly skillful at golf."

In the evening, a group of Sedalians, fair staff, and the presidential entourage dined in style and substance. Dinner decorations consisted of silver vases of American Beauty Roses, garlands of smilax, and bowls of Lady Taft roses. A grouping of flags accented the almond dishes at each plate. The menu included bouillon, creamed sweetbreads and mushrooms served with crackers, turkey with oyster dressing, whipped potatoes, creamed cauliflower, cranberries, cold baked ham, green salad, hot biscuits, olives, celery, cheese balls, and dessert of bisque ice cream, individual cakes, Roquefort cheese, water wafers, and coffee, followed by cigars. At 8 p.m. the president mounted his Pullman car and departed.

Taft spoke at the Coliseum. (V. B.)

Bottom: Missouri mules parade for Taft. (P. C. H. S.)

"The Man from Independence": Harry S Truman

Harry S Truman's 1955 visit to the state fair was not as elaborate as Taft's, perhaps because Truman was not currently president, but also because he was a more down-to-earth man. He came with a smaller group, made a shorter speech, and received much less press coverage. The *Sedalia Democrat* said Truman was "his jovial self and talked as usual in Missouri language to a Missouri audience."

Missourian Harry S Truman had become president at the death of Franklin D. Roosevelt in April 1945 and served until 1952. Truman had been active in Missouri politics prior to becoming president, having served as a District Judge of the County Court (Commissioner), as Presiding Judge (Commissioner), as U. S. Senator, and as the Vice-President. He had visited Sedalia many times, using Sedalia as his campaign headquarters in 1940 in order to distance himself from the corrupt Pendergast machine in Kansas City, and making whistle stop speeches at railroad stations during his campaign of 1948. While a senator, he and his mother had visited the Missouri State Fair.

Harry S Truman admires a Missouri mule. (T. M. L.)

Louis Monsees rides one of his mules on a wagon; Mrs. Monsees holds the bridle. (P. C. H. S.)

In 1955, Truman, then retired to his home in Independence, attended the Missouri State Fair, and spoke at the fifth annual Ham Breakfast held at the Smith-Cotton High School Cafeteria. The head table seated Truman; Senator Stuart Symington; Lieutenant Governor James Blair; Morris Burger, producer of the grand championship ham; Mrs. T. Edwin Bierl, winner of the reserve championship ham ribbon; Missouri Commissioner of Agriculture L. C. Carpenter; Fair Secretary W. H. "Chubby" Ritzenthaler; Judge Frank Monroe; Sedalia Mayor Julian Bagby; Sedalia business leader Walter Cramer; Sedalia

Chamber of Commerce President Kenneth U. Love; the Reverend Ralph Emerson Hurd; and Toastmaster Robert E. Lee Hill. The Sedalia Chamber of Commerce donated the breakfast ham.

Lieutenant Governor Blair and Senator Symington introduced the former president. Truman thanked the men for their brevity, then asked the guests to take off their jackets, as the day was very hot. He took off his

jacket and began to speak. He praised Senators Symington and Hennings for their "absolute harmony" and their youthful enthusiasm for political careers. He quite naturally praised Missouri and the state's agricultural products, noting that his mother had been mortified when in 1935 a Kansas mule won first prize at the Missouri State Fair. "No other hams," Truman claimed, "are anywhere as good as our good old Missouri hams." When he presented twenty-year-old Morris Burger of California, Missouri, with the Honorable Thomas C. Hennings Jr. Trophy for his championship ham, Truman encouraged the increased interest of the young in agriculture.

More than six hundred people attended the breakfast. Truman watched intently as auctioneer Ed Caldwell of Perry, Missouri, assisted by local auctioneers Olen Downs and Jesse Paul, sold the prize-winning hams and presented the money to the Truman Library in Independence. The state fair then presented Truman with a ham, as did Ralls County. The dignitaries then went to the fair, where Truman visited the exhibition and watched horse races.

Prize-winning hams are a delicious part of Missouri's pork industry. (M. S. A.)

President Ronald Reagan admires a young woman's animal. (S. D.)

High Security and Angry Farmers: Ronald Reagan

In 1984, President Ronald Reagan attended the Missouri State Fair. The extensive preparations made for Taft's visit were dwarfed by the amount of time, energy, and work involved in arranging Reagan's trip. Security, the president's safety, and secrecy had become primary concerns since Reagan had been the victim of an assassination attempt in March 1981. The Secret Service wouldn't, for instance, tell the *Sedalia Democrat* how many agents accompanied the president, but two buses were needed to bring them here. Secret Service agents and canine units examined the various highways the presidential motorcade might take from Whiteman Air Force Base at Knob Noster, Missouri, where the president's plane would land, to the fairgrounds. Agents checked culverts, bridges, and fairgrounds buildings for explosives, arranged for "just in case" emergency care at Bothwell Regional Health Center, and set up classrooms at State Fair Community College for press-rooms. Thirty-four extra Highway Patrol officers, including a helicopter unit, the Pettis and Benton Counties' Sheriffs Departments, and the Sedalia Police Department officers assisted the Secret Service. Many Missourians who had lined Highway 50 and Thompson Boulevard awaiting a glimpse of the president were disappointed when the motorcade deviated from the published route and arrived at the fairgrounds via Route Y.

Reagan briefly toured the Shorthorn Barn and saw the prize-winning livestock raised by Missouri youth, then moved to the Coliseum. The Jefferson City High School Jazz Band played "Hail to the Chief" and Missouri Governor Christopher Bond introduced the president. Reagan's speech, praised by the staunchly Republican editor of the *Sedalia Democrat,* was "rich with patriotic

President Reagan visited the fair in 1984. (M. S. F.)

allusions." Reagan lauded America's greatness and declared the nation's intent to "move forward together. And we will do it by strengthening one of our most cherished, vital institutions—the American family farm." As Reagan continued to outline what his administration had done to help the farm economy, an especially touchy issue during that election year marred by the deepening crisis in the agricultural economy, some farmers in the crowd jeered. Another group of farmers, joined by laborers and those opposing U.S. military intervention in Central America, picketed outside the Coliseum, claiming that Reagan's policies "broke the farmers' back."

In addition to making a speech in the Coliseum, the president spent thirty minutes meeting in the Youth Building with the Governor's Advisory Council on Agriculture and sixty agricultural and political leaders. Governor Christopher Bond and Department of Agriculture Director James Boillot arranged the meeting. During the meeting Representative Ike Skelton spoke on behalf of Missouri farmers when he said, "I hope the president used the trip to learn that the economic recovery bypassed rural America. Past actions have demonstrated to me that his administration has written off the rural American farmer."

The Fastest Horse in the World—Dan Patch

Fast horses always appear at the fair, but on October 4, 1909, Dan Patch, the world's fastest horse, ran an exhibition race against the "world's second fastest horse," Minor Heir. At the time, a very good pacer might sell for $2,000; Marion W. Savage had purchased Dan Patch in 1902 for $60,000 and Minor Heir in 1909 for $10,000. Patch had set several records, winning fifty-four out of his fifty-six races. Dan Patch's record, set in 1905, was 1:55; Minor Heir's best time was $1:59 1/2$.

Dan Patch raced against Minor Heir in 1909. (V. B.)

Savage's horses were well known in part because he marketed them aggressively, using them as advertising icons for items as varied as tobacco, automobiles, International Stock Food animal feeds, and featuring pictures of Dan Patch on picture cards and thermometers. Savage took a percentage of gate receipts rather than a fixed fee for staging the races. The Sedalia press bragged that "an immense crowd" of thirty thousand people watched the race. Savage, according to the minutes of the Executive

At the "Dan Patch" Race, State Fair, Sedalia, Mo.

Committee of the State Fair Board of Directors, was to be paid sixty-five per cent of the amount Monday's grandstand and gate receipts totaled above the average of receipts from the 1907 and 1908 Monday attendance.

Harry Hersey, the horse's long-time trainer, drove Dan Patch, while popular Sedalia horseman Bill Turner drove Minor Heir. The *Sedalia Democrat Sentinel* noted that the horses had previously "appeared only on very bad half-mile tracks" and predicted "great interest" in a race on a "good mile track." The race was exciting, but no records were set. Dan Patch's time for the mile was 2:07, and Minor Heir's was 2:07 1/4. Savage made a motion picture of the race, so that "it may be handed down to posterity."

The Missouri Centennial Commission poses in the state capitol. (M. S. A.)

Below: This poem appeared in the *Program Book,* Missouri State Fair, 1921

Special Celebrations 1921—One Hundred Years of Missouri Statehood

Missouri entered the union in 1821, just seventeen years after the Louisiana Territory was purchased. The state, which had once been home to several Native American nations, came into the union as a result of the compromise of 1820, which allowed Missouri to enter as a slave state while Maine entered as a free state, thus preserving the balance of slave and free states. These aspects of

A Golden Smile and the Folks Worth While

In Honor of Missouri's 100th Birthday

BY WILL FERRELL

A hundred years is a long, long time
To weave into meter and lore and rhyme,
So, briefly, I'll sing of a golden smile
And a plea to the hearts of the folks worth while.

Born at the dawn of a distant day,
A babe as fair as a rose in May.
Rocked in a cradle of "hick'ry split"
By the proud young parents, Faith and Grit.
The babe was swaddled in homely things
And fed from a bottle of living springs.
Schooled in the ethics of tree and sod,
She grew to youth in the land of God.

She wasn't pampered in idle dreams,
Her world was real with its hills and streams;
Where the wolf stalked forth when the sun was gone
And the "bob-cat" challenged the crimson dawn;
Where the "pot hound" blinked at the old "smooth bore"
And a coon-skin hung on the cabin door;
Where the grist mill thundered its awkard lay
To the simple folk of an early day.

Where "Sorghum Holler" and "Possum Run"
Were the rainbow's end to a 'bar'l o'fun."
Here the "gal growed up" as gals will do,

Nurtured in toil when the land was new.
All "fussed up" in a homespun gown
For a "mule-back" ride to a cross-roads town.
Comely and shy as the piping quail
In the old worm fence by the river trail.

Her fame soon grew as a winsome Miss
And suitors fought for her golden kiss.
But this is the edict she spake to all:
"My charms are gratis to great and small.
All ye who would strive for my golden smile
Must measure up to the folks worth while."
And, speaking thus, she spread her charm
To the swains of the village and field and farm.

That's how it started. A hundred years
Since the babe was born to a world of fears.
Through calm and tempest, through sun and rain,
She breathes her message to hill and plain;
For fair Missouri retains her youth
And her early precepts of love and truth.
And, true to her standards, her lure, her smile,
We're glad to be one of her "folks worth while."

Sorghum Holler and Possum Run
Are still extant in the realm o' fun.
But the swaddled babe is a woman grown.
With a sizeable family, all her own.
She rides no more to the "deestrict skule"
In a homespun frock—on a flea-bit mule,
But she goes "de luxe," like a millionaire
To greet her kin at the big State Fair.

Missouri history would figure prominently in the fair's celebration.

In 1921, Missouri observed its centennial. Events commemorating statehood occurred throughout the year, but the largest celebration happened at the Missouri State Fair. "Mother Missouri" called her "children" home and hosted a thirteen-day extravaganza in Sedalia. The celebration included production of a pageant showcasing the history of the state, special exhibits of Missouri products, the initial distribution of a commemorative half dollar, and the appearance of a number of prominent Missourians. Fair supporters expected one million visitors, including President Warren G. Harding.

The Pageant of Missouri, written and directed by George H. Hoskyn, featured five thousand performers on a stage six hundred feet wide and seventy-five feet deep. Alexander de Beers, Harry Russell, and Homer Goulet designed and painted the forty-five-foot-high scenery. The *Souvenir Program* described the stage as the "largest outdoor stage ever built for a production of this sort."

A postcard advertises the Missouri Centennial and the State Fair. (V. B.)

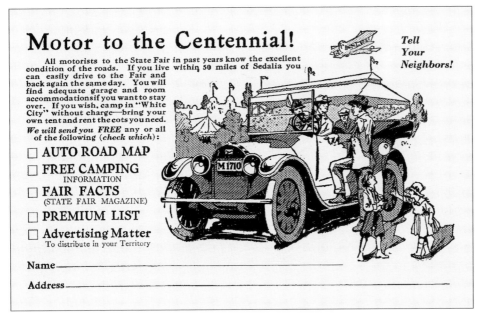

Producer J. Saunders Gordon and General Manager and Scenic Producer Edgar Larmer worked with electrician William H. Erwin, assistant electrician Frank Hoskyn, property master Harry Hoskyn, and wardrobe mistress Mrs. E. A. Wood. Ballet Mistress Virna Herman Walker and Assistant Ballet Mistress Mae de Beats managed Premiere Dansuese Hazel Wallack, the Wallack Ballet Company, and the thirty-six local children who danced as Flora Fairies and Spirit Fairies in the Children's Ballet.

Below and facing page: The U.S. Mint created a half-dollar in honor of Missouri's centennial. (V. B.)

Master of Ceremonies George H. Hoskyn and his assistant Harry Wallace introduced scenes from Missouri history dating to the earliest European exploration in the mid-sixteenth century. The *Souvenir Program* featured historic background information written by Floyd Shoemaker, secretary of the State Historical Society of Missouri and editor of the *Missouri Historical Review* and illustrated by Chapin. The Pageant, divided into three periods—Discovery, Development, and Achievement—featured reenactments of historic events such as the meeting of the Osage Indians and Father Marquette and Joliet, Lewis and Clark's explorations, Civil War battles, and the return of soldiers from the Great War. In addition, ladies draped in Grecian gowns portrayed the mythic qualities of Agriculture, Metal, Learning, and Industry. Named Agricola, Mettalla, Scholastica, and Industria, they were attended by other ladies draped in similar gowns.

The Missouri National Guard under the command of Adjutant General W. A. Raupp played the roles of soldiers in the Colonial, British, French, Union, and Confederate Armies. The Sedalia Chamber of Commerce organized the remaining actors. Thirty-one women, forty-five men, and twenty-three children played pioneers, twenty men played trappers, twenty men played steamboat men, twenty men played rivermen, and twenty men played trainmen. One-hundred fourteen women represented each of Missouri's counties. A group of African American dancers participated in various scenes.

The issue of a special centennial half dollar marked the 1921 fair. Robert Aiken designed the coin, which featured on its face an image of Daniel Boone in coonskin cap and buckskin shirt. The dates 1821 and 1921 flank the portrait, and the words *UNITED STATES OF AMERICA* and *HALF DOLLAR* circle the coin. The reverse side of the coin shows Boone and a Native American standing surrounded by twenty-four stars, signifying Missouri's admission to the Union as the twenty-fourth state. *MISSOURI CENTENNIAL* tops the image and *SEDALIA* underscores it. Five thousand of the coins were marked with *2H4* and sold as collector's items. Today, these coins are highly prized by collectors.

The Centennial Celebration drew 287,970 people to the fair, far less that the one million expected, and without President Harding. The attendance that year, though slightly lower than that of the 1919 and 1920 fairs, was considerably higher than that of later years. The Centennial Celebration, lauded as

The Program for the 1921 Fair included the *Pageant of Missouri*. (M. S. F.)

"the most important gathering of Missouri citizens since the founding of the state," was costly. Early in 1922, the Attorney General noted that the fair that year had spent the money appropriated to the fair by the General Assembly for both 1921 and 1922, plus all receipts from 1921, and was still $21,000 in debt. He issued a terse warning to the board of directors: "If you hold a State Fair," you must make a "carefully considered budget showing all the expenses you will incur . . . and endeavor to hold the total expense of the State Fair to within or less than the anticipated revenues." The Board members were concerned, as they felt they must continue to hold a fair despite budget constraints. Secretary Bylander resigned, and W. D. Smith became secretary. Within a year and with careful management, the state fair was solvent again.

The 1976 Program Guide featured a flag-like design. (M. S. F.)

Two Hundred Years of the United States—1976

The nation's bicentennial, held in 1976, commemorated the signing of the Declaration of Independence. Cecil Owen, coordinator of bicentennial events, had planned a best beard contest, a look alike contest for people who thought they resembled historical characters, and an old-time fiddler's contest. Crafts items such as red, white, and blue knitted items, decoupaged historical collages, pressed flowers in wooden frames, and apple dolls emphasized bicentennial themes. Horticulture exhibits by Garden Clubs featured "freedom" as a theme in flower arrangements. "We've been planning these events for a long time and now the time has finally come. We're all looking forward to seeing the entrants. We're hoping to get quite a few," Owen commented as the fair opened. "Would-be Betsy Rosses and Uncle Sams" crowded the grounds, and visitors listened to "old-time band music" in the grandstand. Tex Beneke and Ray Eberlee and the Modernaires with Paula Kelley recreated big band sounds with emphasis on Glenn Miller hits.

Other events included the display of an "authentically reconstructed pioneer home." Owen and Mrs. Jim Mathewson had brought a dismantled log house from Benton County and rebuilt it on the fairgrounds. Mrs. Peggy Hale and Mrs. Marilyn Hunnell served as hostesses in the house furnished with antiques, including a rope-spring bed, a tin-headed doll, a broad

Rock and Roll!

In 1974, the Ozark Music Festival hit the Missouri State Fairgrounds with bands, amplifiers, thousands of people, and tons of trash. Marketed to Sedalia and the Director of Agriculture as a family-oriented bluegrass festival, the event was publicized throughout the country as a rock concert similar to Woodstock, with performances by Lynyrd Skynyrd, the Nitty Gritty Dirt Band, the Charlie Daniels Band, the Eagles, the Ozark Mountain Daredevils, Bob Seger, and America. Despite the warnings of Sheriff Emmet Fairfax that the festival promoters could not provide adequate security, plans for the festival continued. On July 19, 180,000 people descended on Sedalia, blocking all lanes of Highway 65 as they attempted to enter through the one open gate of the fairgrounds. The weather, typical of Missouri in July, was hot. Concert promoters had not planned for an adequate first aid facility, nor for showers, toilets, ice, water, or food. Dr. A. J. Campbell worked valiantly to maintain emergency health care, but was overwhelmed by the number of victims of drug overdose, heat stroke, and injuries. Water lines were shut off on Saturday afternoon, intensifying problems. Motorcycle gangs took over the campgrounds area. According to researcher Anneliese Homan, Wolfman Jack, who had advertised the festival on his radio broadcast, lamented the out-of-control nature of the crowd. On Sunday, July 21, National Guard members came on the fairgrounds to bring in medical help and supplies.

When the festival was over on Sunday, the crowd drifted away. The State of Missouri was stuck cleaning up the garbage and repairing the damage to the fairgrounds. A helicopter sprayed lime, bulldozers removed topsoil, and truck after truck hauled garbage to the landfill. With a mammoth effort, state fair crews had the grounds ready for the opening of the state fair a month later.

The Ozark Music Festival brought thousands of young people to the fairgrounds. (S. B.)

ax, a four-pronged hay reaper, coal oil lanterns, a wooden flour bin, quilts, and a butter churn. Herb Templeton of Arrow Rock demonstrated blacksmith work and Ralph Levings showed a rebuilt steam engine.

The Missouri National Guard presented three different shows. The First Missouri Volunteers from the 139th Air Refueling Group of St. Joseph, dressed in military uniforms appropriate to the periods they represented, displayed the flags flown during ten important periods in American history. A second show involved The Governor's Own, reenacting Washington's Virginia Militia, carrying Brown Bess rifles and performing drills used by Continental Army. The third show, the "Yankee

Doodle Dandy Puppet Show," delighted adults and children alike with its humorous depiction of events from colonial times, including the first Thanksgiving and the creation of the Stars and Stripes.

State Fair Anniversaries—Twenty-Five, Fifty, and Seventy-five

In 1925, the Missouri State Fair celebrated again; fair officials announced in July the "plans for an impressive dedication ceremony commemorating the silver anniversary." Several buildings on the grounds were repainted for the occasion, extra flowers had been planted and were in full bloom, and flags and bunting decorated the buildings. Tri-colored lights made the grounds at night look like a "miniature fairy land"; the show was valued, according to the *Sedalia Democrat*, at five million dollars. Secretary Smith proclaimed, "There are a greater number of exhibits than ever before in my administration. I'm looking forward to a wonderful fair."

The one-hundred member Boy Scout Band from Springfield, the M.K.&T. Band, the 175 member Henry County Band, the Missouri Pacific Band, and other bands presented concerts on the grounds. Three carloads of dignitaries, twenty-three floats, and seventeen hundred Katy Railroad employees from Parsons, Kansas, accompanied four bands—the Katy Band, the Parsons' Junior Band, the Franklin Band and the Sedalia Band—as they paraded through downtown Sedalia. Governor Sam A. Baker dedicated the fair in a speech before a crowded grandstand. Colonel Matthew Pagelow of Scott Field, Belleville, Illinois, flew a dirigible to Sedalia. Actress Dorothy Dawn, the former Sedalia girl Dorothy Ilgenfritz, appeared with the mayor of Los

Angeles, and the Loos Shows provided entertainment on the Midway. Fifteen Native Americans, including several children, came to Sedalia to appear in Hal Worth's pageant about the settlement of the American colonies.

1952—The Golden Anniversary

In 1952, the Missouri State Fair observed its golden anniversary. The fair opened on August 17 with a record opening day attendance of 25,055. The fireworks exhibit titled the "Gay 90s Review" featured fireworks representing a cake with candles, a bicycle built for two, and a surrey with fringe on top as part of special fireworks exhibit. Rodney Polson, assistant director of the University of Missouri band, directed thirteen hundred high school band students in parades through the grounds each day and massed concerts each evening.

A clown advertised the 1952 fair. (M. S. A.)

The Pettis County Historical Society sponsored a square dancing exhibition, singing, dancing, skits, piano music, and an exhibit of local artifacts and oddities. The Floriculture Department offered an exhibit of wedding flowers from 1901 through 1952.

Governor Forrest Smith visited on Wednesday and went up in the Stag Blimp. Smith reviewed his term, expressing pride in the ten-year road program, a driver's license law, a balanced budget each year, more state aid to schools, increased assistance to the elderly, the disabled, dependent children, and the blind, but admitted disappointment in having failed to reduce the state income tax. When the Golden Anniversary Stage Show and Musical Review in the grandstand Wednesday evening was delayed by the completion of the Grand Circuit horse races, Cowboy Bob Atcher, "foreman of the Meadow Gold TV Ranch" a television show sponsored by Beatrice Foods Company, rescued the show by coming on stage and singing and playing his guitar. Atcher, a very popular entertainer "known to millions of television and radio fans," delighted the visitors.

A tram car advertised the dates of the 1952 fair. (M. S. A.)

The Elkettes, a drill team from Holdenville, Oklahoma, marched in 1952. (M. S. A.)

During the golden anniversary year, the weather demonstrated its complete power over fairgoers. At 1:20 Thursday morning, August 21, 1952, a tornado raged through western Sedalia, its "wrath centered" on the Midway. Eye-witnesses cited by the *Sedalia Democrat* reported the storm "struck the Swine Pavilion, slammed down the south side of the carnival midway, veered to the northward at the Cetlin-Wilson gate, turned to the east near the Missouri-Pacific gate, banged through the cattle and horse barn section, slid southeastward to the grandstand area, ripped across the race track, then left the fairgrounds in a northeasterly path." The storm leveled the Veterinary Building to the east of the Midway and destroyed the $500,000 exhibit of tropical butterflies and moths in the building. Twenty of the Midway's twenty-five rides were damaged, and light poles downed. A jewelry stand was carried one-half mile from the yard of the Administration Building to the southern fence near the M.K.& T tracks. The storm hurled the Missouri-Pacific Railroad office building into the air and down on a tent housing Milking Shorthorns, killing a prize-winning bull. Trees fell onto cars and trucks. The bandstand lost its roof, and the racetrack was flooded. The storm picked up a house trailer occupied by Harry Lee Pyle, his wife Virginia, son Thomas, and daughter Jeanne Ann, carried it sixty feet, and turned it on

The 1952 tornado flooded the racetrack, damaged the Midway, and toppled trees. (M. S. A.)

its side. Pyle was killed, and his wife and son injured. Howard McKinley, Louisa McKinley, William Cecil Goldey, Albert William Slaughter, Levi Taylor, Elizabeth Copple, Ray Junior Burd, William Roland, James Wiley, and Curtis Gale were also injured.

Red Cross workers arrived at 4 a.m. and served coffee, rolls and doughnuts, milk, ham sandwiches, and cheese sandwiches to carnival workers, campers, and fair personnel. Cleanup began at dawn. Rain continued to fall as stock owners moved their animals, tents were reset, and stands cleaned. Sedalia police officers directed traffic away from the fairgrounds to enable repair crews easier access. Cooperation among the "several thousand city, fair, and carnival employees…worked a miracle clearing up tornado debris," bragged the *Sedalia Democrat* on August 22. The fair opened again Thursday evening. Nearly three-quarters of the Midway rides and shows had been "put back in working order," though the badly damaged Ferris Wheel had been sent to its factory for repairs. Commissioner of Agriculture Robert Thornburg praised the concessionaires, exhibitors, carnival workers, and fire and police departments for their ability to work together to allow the fair to continue: "It has been gratifying the way everyone cooperated."

A group of entertainers provided an impromptu show to replace the grandstand horse shows that had been rained out. That evening, eight thousand people saw the Missouri State Hall of Fame Athletic Award ceremony and the Golden Anniversary State Show. On Friday morning, Girl Scouts took damaged stuffed toys from the Midway to the Girl Scout building where they were cleaned and repaired to be given to needy children. Although harness races were cancelled for the week, the "big car automobile races" ran Saturday afternoon and stock cars races ran Sunday. Truly, the show did go on.

This undated photo shows a couple celebrating their fiftieth anniversary at the fair.

1977—The Seventy-fifth Anniversary

In 1977, the fair was seventy-five years old. It "opened Friday morning as fresh and frisky as the prize show horses" according to *Sedalia Democrat* reporter Ron Jennings. "The popular three C's—corn dogs, cattle, and the carnival—will continue to highlight key traditional attractions. A fourth traditional C, crowds, are also expected despite drought conditions in some parts of the state." The fair reflected issues of the day—awareness of the needs of handicapped people, recognition of senior citizens, energy conservation, and the importance of agriculture to Missouri's economy. Governor Joe Teasdale praised Handicapped Day and the twenty-four hundred "handicapped citizens receiving the dignity they deserve" during a visit to the fair. Senior citizens participated in the festivities; the twenty-three piece Kitchen Klatter Band from Belton performed old songs such as "I've Been Working on the Railroad." Senior citizens competed in a contest recognizing the best performance of "The Missouri Waltz." An exhibit by engineering student Loyd Wright of the University of Missouri at Rolla encouraging people to "produce and conserve all available sources of energy" included a "blinking, brightly-lit computer called an 'energy-environment simulator'." Teasdale recognized the "young people of our state's farming families working very hard" at the fair and noted that Missouri agriculture was beginning "a new era of being promoted abroad in terms of foreign markets." The prize winning Missouri ham sold for a record breaking $4,000; Jerry Murphy of Murphy Brothers Shows bought the ham and presented it to Governor Teasdale. Competitions that year included an auctioneering contest, a Missouri wine category, and a World Steer Show with $1,500 first place and $1,000 second place prizes. The fair offered $3,633,885.50 in premiums overall. 307,562 people attended the fair described by Jennings as "full of the spunk and spirit that has made her one of the outstanding state fairs in the country."

Governor Joe Teasdale visited the fair in 1976 and 1977. (M. S. A.)

The State Fair Goes to War

In April 1917, the United States entered the Great War, fighting against the Triple Alliance of Germany, Austria-Hungary, and Italy. The war influenced activities at the fair, calling forth an increase in patriotic activity. The 1917 fair motto recognized the war effort: "It Promotes Patriotism, Production and Preparedness." The *Souvenir Missouri State Fair 1917* booklet reminded visitors, "Every energy of the 1917 Missouri State Fair has been exerted towards educational work that would lead to increased production at reduced costs for Missouri farmers. Governor Frederick F. Gardner, himself a keen business man, realizing the importance of the work being done by the Missouri State Fair for increased production, as well as the highly patriotic tendencies of all its work, with a view to ultimate preparedness, issued a proclamation on June 1 asking all Missourians to give the Missouri State Fair their hearty and unqualified support. . . . Patriotic fervor in the displays of military and naval equipment, together with the display of the Red Cross, is promised." The *Program* promised a free vaudeville show of five acts between the races and an "automobile fashion show" under the grandstand.

One patriotic exhibit highlighted the activities of the Missouri Council of Defense. William F. Saunders, secretary of the Council of Defense wrote that the exhibit of ten placards and posters from other states and five hundred large war photos given by the National Security League of St. Louis had been "seen by thousands." The exhibit also included a display of the work of the Women's Council of Defense arranged by Mrs. Olive Swann, organizational charts, National Safety Council charts "dealing graphically with the conservation of life and limb," and the war poster campaign of the Missouri Pacific Railroad.

The 1918 fair continued the emphasis of patriotism. Thursday, August 15, was designated Patriotic Day, with activities directed by the Council of Defense. The federal government had prepared a war exhibit circulated among various state fairs in 1918. The exhibit showed a vast array of war equipment, including weapons, signal corps carrier pigeons, portable field wireless outfits, aviator's clothing, airplane machine guns, gas masks, uniforms, campaign badges and medals, trench periscopes, charts showing increases in personnel, Navy mines, depth bombs, Marine Corps equipment, captured German rifles, and relics from the French, Italian, British, and Canadian forces. The Department of Agriculture showed "propaganda material of the educational division" regarding food conservation and food substitutes, and the Department of Commerce displayed its experiments in the use of fish skin as substitute for leather. "Designed to be a war show from first to last," according to the U.S. Department of Agriculture, the exhibit also displayed replicas of "a Y.M.C.A hut, similar to those that radiate cheer on the western front, a Salvation Army dugout like those that

A military company marched in a parade in 1916. (V. B.)

No. 33. AMERICA WE LOVE YOU
Fitted in OUR CELEBRATED GILT PAPIER MACHE
FRAMES. Size Outside 20 x 24 inches.

No. 40. COLORED MAN IS NO SLACKER.
LOOKS FINE IN PAPIER MACHE FRAME.

The Council of Defense
exhibited war posters in 1917
and 1918. (M. S. A.)

furnish doughnuts and coffee to the boys on the firing line, and an exhibit of
Red Cross work, from hospital to canteen activities."

The Missouri Council of Defense sponsored demonstrations of cooking,
canning, and dehydrating, with an emphasis on recipes for "war bread."
Exhibits about clothing conservation, textile buying, women's work in agriculture, and raising war gardens targeted the need to conserve and produce
resources at home, while displays of knitting and making surgical dressings
pointed out work Missourians could do to benefit the men on the front. The
Red Cross signed up volunteers. The Sedalia Department of Child Welfare
sold the booklet "Mother Goose in War Time" to raise money for its work.

World War II

World War II had begun in Europe in the late 1930s with the annexation
by Germany of Austria and Czechoslovakia and in Asia with the Sino-
Japanese war. The United States entered the war against the Axis powers of
Germany, Italy, and Japan in December 1941, following the Japanese attack

Canning clubs emphasized at-home production and preservation of food. (M. S. A.)

Below, right: These Ferris wheel riders at the 1941 fair seem unconcerned with the approach of war. (V. B.)

on Pearl Harbor. All aspects of American life were changed by the war, and those changes were reflected in the Missouri State Fair. Although fairs during the 1930s had recognized preparedness for war as hostilities escalated in Europe, the fair in 1942 dedicated itself "to American Victory through the advancement of Agriculture, Industry, Commerce, Art and Science."

War related exhibits dominated the 1942 fair. The Office of Civilian Defense exhibit included information on firefighting, on price controls, and on scrap collecting. Visitors viewed films about the war effort in the Education Building. Fort Leonard Wood soldiers demonstrated many aspects of military activity—how to build a water tower and purification plant, how to use a hectograph to duplicate maps, how to camouflage trucks and pup tents, how to run wire entanglements, and how to build approach spans for light pontoon bridges and fixed bridges. Visitors saw military equipment such as assault boats, personnel carriers, and bombs.

Since most adult men were away fighting, the Boy Scouts camped on the grounds and worked as parkers, ticket takers, ushers, and messengers. The Girl Scouts' exhibit, located in the Coliseum, focused on war preparedness. Canned fruits and vegetables dominated the homemaking category. A doll showing proper first aid treatment for assorted injuries attracted a great deal of attention.

The Works Progress Administration Art Project prepared an exhibit of posters about civilian wartime activities for the State Council of Defense. Hugh Stephens of the Missouri State Council of Defense called on Missourians "to contribute something everyday to the war effort, to help and uphold the hands of civilian leaders charged with the responsibility of providing all air, fire, police, and medical protection which the people of this country can hope for under the attack of war." New Air Raid Wardens and Auxiliary Firemen received certificates at graduation

exercises on August 26.

The regular fair departments supported the war effort as well. The Home Economics Department recognized the importance of conservation of food by requiring that foods entered use at least one-third honey as replacement for rationed sugar. A State Fair Cooking School, held in a tent behind the Administration Building, taught how to cook with wartime rationing. Each day a full meal was prepared and given to audience members. Victory garden exhibits encouraged the growing of "practical foods that add vitamins to the diet." As part of the focus on "supplying a family in wartime," the judge, Dr. A. D. Hibbard of the University of Missouri, emphasized canning or drying of foods in ways that preserved nutrients. Another aspect of civilian defense involved the over three thousand livestock entries demonstrating the most economical ways to produce meat and dairy products.

A group of men recently inducted into the military march at the 1946 state fair. (M. S. A.)

Below: The Missouri State Fair resumed in 1945. (V. B.)

The "Mercy Angels Pageant," written by Thomas Wolff of St. Louis and presented by the Pettis County Red Cross on Wednesday night, depicted the work of the different departments of the Red Cross in a series of tableaus. Mrs. Frank S. Leach and Robert A. Drohlich directed the pageant, narrated by Eunice Cousley and Al Drohlich. Fifteen scenes showed Florence Nightingale, Clara Barton, a refugee camp, a blood transfusion, wounded soldiers, a Red Cross recreation area, and Red Cross nurses. Phil McLaughlin, chairman of the Pettis County Red Cross, and Mrs. Arthur Kahn, chair of the women's division, organized the 125 actors. Missouri National Guardsmen played the roles of soldiers.

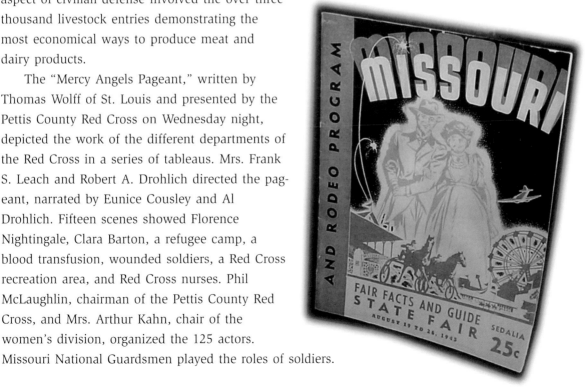

The total attendance for the 1942 fair was 146,750, somewhat lower than attendance in 1940 and 1941. As resources, including food, shoes, gasoline, and tires, were rationed, unnecessary travel discouraged, and the war effort intensified, the board decided not to hold a fair in 1943 and 1944.

By May 1945, victory in Europe had been declared, though the war still raged in Japan. The Department of Agriculture and the General Assembly had discussed not holding a fair that year, but recognized a clause in the contract between the state and the Van Riper family, who had donated the land that Sedalia offered to the State Fair: If a fair were not held for three consecutive years, the land would revert to the Van Ripers. John Ellis, Commissioner of Agriculture, wrote, "In order to safeguard the title to the one and one-half Million Dollar State Fair Grounds and buildings, it is necessary to hold the 1945 State Fair. The winning of the war in Europe will permit a return of many peacetime activities which will increase the interest in fairs." The fair would go on. Senator Frank P. Briggs encouraged Missourians to attend the fair: "They should gather there . . . to recovenant with themselves that they will leave no stone unturned to produce to the fullest, whether in the office, in the shop, in the factory, or on the farm, that those boys of ours who are still meeting resistance on our campaign for the cause of Democracy, may be accorded the fullest support of every kind."

The Korean Conflict, sometimes called "America's forgotten war," had little impact on the fair. The Vietnam war, though our country's longest and most costly war, also had little effect. The real impact of the Vietnam conflict came in 1997, when a replica of the Vietnam Veterans Memorial came to the fair. The memorial wall, designed by Maya Lin, lists the names of the fifty-eight thousand who died during the conflict. Vietnam veteran and Pettis County Sheriff Gary Starke participated in the dedication of "The Wall," praising those who served as "the best and brightest of America. . . . They wanted to come home to their loved ones, home to their communities,

A robotic arm applies lipstick in this Atoms for Peace exhibit. (M. S. A.)

Right: The Gemini spacecraft appeared at the fair. (M. S. A.)

home to the country that they loved and served. . . . Their memories and sacrifice are enshrined in our hearts forever."

The Cold War and the space race also made an appearance at the fair. During the late 1950s and early 1960s, the nation was increasing its defense capabilities and the Air Force began the process of installing Minuteman missiles. Whiteman Air Force Base in nearby Knob Noster, designated a "support base" of the Minuteman ICBM, hosted a display of a full-size model of a Minuteman missile and its transporter-erector at the 1964 fair. Other Cold War era exhibits included the "Atoms for Peace" exhibit, touting the safe uses of nuclear power and of robotics, and a NASA Space Capsule.

Economic Difficulties

The fair's economic status reflects the prosperity of the state and the nation, as well as the generosity of the General Assembly. In the early years of the fair, the General Assembly was reluctant to approve appropriations for the fair and the grounds, but eventually provided for the construction of buildings "essential to a proper display of the unlimited resources of the state" that the State Industrial Association had called for in December 1900. Secretary W. D. Smith heeded the Attorney General's warning about staying within a careful budget following the overspending of 1921, and brought the fair back to economic stability. National events would cause the fair's next major economic crisis.

The stock market crash of October 1929 plunged the country into the Great Depression. Missouri had already experienced economic difficulties as agricultural prices fluctuated and taxes on land rose during the 1920s;

Top: The Vietnam Memorial Wall generated a mixture of emotions. (S. B./S. D.)

The Alliance for Progress encouraged peaceful relations between countries of the world. (M. S. A.)

The Merci Train

Visitors line up to see the Merci Train car. (M. S. A.)

In 1948, Americans, led by columnist Drew Pearson, sent food and relief supplies to France, still devastated by World War II. The French responded in 1949 by sending a "Merci Train" in thanks. The French loaded forty-nine boxcars, one for each state plus one to be shared by Washington, D.C. and Hawaii, with gifts. The boxcars, called "40 and 8 cars" had been built between 1872 and 1885, and were used in World War I and World War II to transport either forty men or eight horses. The boxcar at the Missouri State Fair was brought from the Jefferson City railroad yards with the help of the Pettis County Commission and La Societe des Quarante Hommes at Huit Chevaux, popularly called the Forty and Eight, an organization of veterans and part of the American Legion. In 1950, according to Forty and Eight member Lawrence Roe, the car was placed on the fairgrounds, originally set across from the Womans Building and later moved to its current location. Grand Chef de Gare H. B. Herider of the local Forty and Eight reproduced the original plaques, representing the provinces of France.

A delegation from St. Louis visited the fair in 1937. (M. S. A.)

Below: The fair continued through the dark days of the Depression. (V. B.)

however, the combination of severe drought and falling prices during the early 1930s caused farm income and land values to decline further. Historian Richard Kirkendall sees a connection between unemployment, falling prices, and declining land values, noting that the "index of prices paid to farmers dropped...from 146 in 1929 to 65 in 1932" and that the price of "farm land dropped from $53.23 per acre in 1930 to $31.36 by 1935." Unemployment increased "to above thirty-eight per cent in 1932 and 1933, well above the national average.

On May 8, 1933, Fair Director W. D. Smith wrote Governor Park suggesting that a fair not be held that year, and that appropriations for the fair be spent instead on repairs to fair buildings: "There is $1,000,000.00 worth of buildings and practically all of them need attention." He further noted that Missouri was participating in the Century of Progress Exposition in Chicago and should focus its attention there. By June 29, Park had replaced Smith with Charles Green. The Sedalia Chamber of Commerce rallied to support the fair, with "Good Will Tourists" traveling to nearby towns to encourage attendance at the fair. At fair time in 1933, the *Sedalia Democrat* editorialized: "Economic conditions nearly brought an end to the State Fair this year. With state funds near the zero mark, it was thought for a time that the institution so dear to many citizens would have to be given up, temporarily at least. But the desire to see the traditions of the fair continue surmounted monetary obstacles and the sponsors and backers of the Missouri State Fair deserve commendation for their splendid, untiring efforts in its behalf." Though attendance fell below the 1931 figure of 225,000, 155,000 visitors attended the 1933 fair, an increase over the 146,250 attendance of 1932.

The 1934 fair indicated the severity with which the Depression, coupled with severe drought, affected Missourians. Months devoted to planting— April, May, and June—had been very dry, and June, July, and August recorded the hottest average summer temperatures on record. Fair attendance fell to 85,250, the lowest since 1917. In an attempt to increase attendance, the fair initiated new policies in 1935. The first lowered admission prices to twenty-five cents for adults. The second drastically reduced the number of passes given out and required everyone to pay admission. These policies, coupled with better crop years and economic recovery, boosted fair attendance to 209,800. By 1937, "Prosperity has returned," announced the *Sedalia Democrat*, "evidenced by the crowds attending the Missouri State Fair, men and women, boys and girls, who come from far and near to view the products of the state and to enjoy the many attractions assembled on the grounds."

Beauty Queens

In 1935, the Miss Missouri pageant began at the state fair. Organized by Mrs. Jo Beasmore in Pettis County and then throughout state, the Missouri pageant would select a representative to compete in the Miss America pageant at Wildwood by the Sea, New Jersey. Edna Smith, a twenty-one-year-old junior at Central Methodist College in Fayette, won. The pageant was filmed by Paramount News for showing at theaters. In 1936, Smith passed her crown on to Margaret Price of Lexington.

In 1958, the fair began its own beauty pageant. The Missouri State Fair

Miss Senior Missouri

In 1997, the state fair hosted a different sort of beauty pageant. Recognizing the beauty that accompanies maturity, the Miss Senior Missouri highlights what Pageant Director Del Hoffman called "competitive spirit, dignity, and ongoing zest for a challenge." Joan Daues of Ellisville, the mother of seven children and grandmother to eleven, was crowned by former queen Jerre Brown. Daues sang "Musetta's Waltz" from *La Bohéme* as part of the competition, and explained her emphasis on achieving goals: "It's not what happens to you, but how you respond to it."

Governor Blair appears with queen candidates. (M. S. A.)

2001 Queen Amanda Dyke shows her second-prize winning hog. (A. D.)

Queen candidates enjoy hot dogs. (M. S. A.)

Right: The newly crowned queen receives a kiss from Thomas Eagleton. (M. S. A.)

Queen, chosen from the winners of County Fair Queen contests, represents
the state fair and its interest in agriculture throughout the state. Contestants
are judged on their talent or public speaking ability, their poise in answering
questions in interviews with the judges, and attractiveness. The winner
receives a $1,000 college scholarship.

Management

When the General Assembly authorized the state fair in 1899, it created a
Board of Directors consisting of one member from each senatorial district,
with the dean of the College of Agriculture at the University of Missouri, the
state superintendent of schools, and the governor as ex-officio members. The
board elected an executive committee and officers who managed the yearly
fair; the secretary was responsible for the daily details of the fair. In 1933,
the General Assembly abolished the Board of Directors. The fair was made a
division under the Department of Agriculture and the position of secretary of
the fair became a political appointment. Under this system, fair leadership
might change each time a new governor was elected. In 1996, the General
Assembly created a State Fair Commission, a group of eight individuals, chosen equally from the two political parties and appointed by the governor,
who work with the Director of Agriculture to oversee the financial management of the fair; one of their duties is to hire the director. This system has
enabled the fair to operate more smoothly.

Directors of the State Fair of Missouri

Not shown:

J. R. Rippey
1901–1907

John T. Stinson
1908–1914

E. T. Major
1915–1916

E. G. Bylander
1917–1921

W. D. Smith
1922–1932

Charles W. Green
1933–1941

Ernest W. Baker
1942–1945

Roy Kemper
1946–1950

William E. Preston
1951

Rollo Singleton
1952–1953

Ross Ewing
1954–1955

W. H. Ritzenthaler
1956
1961–1964

M. C. Ervin
1957–1960

Wilbert Askew
1965–1972

Ronald Jones
1973–1974

Jerry Hermann
1975–1977

Harold Hurd
1978–1979

R. D. Nichols
1980

Marion Lucas
1981–1985

Bill Waddell
1986–1988

Roger Alewel
1989–1993

Bill Arthaud
1994–1996

Gary Slater
1997–2000

Mel Willard
2001–Present

CHAPTER 7

"AGRICULTURE, INDUSTRY, ART AND COMMERCE":

The Best of Missouri for Missouri

Polk County's exhibit emphasizes its grain production. (M. S. A.)

BEFORE THE 1903 FAIR, THE BOARD OF DIRECTORS PRAISED the "great agricultural wealth of the state, the fertility of our soil, the variety and excellence of our products and our favorably climactic conditions." Early fairs were focused almost exclusively on agriculture. Fair historian Wayne Campbell Neely, writing in 1935, saw "the exhibition of agricultural products for prizes . . . the essential feature of the fair." However, he

considered "the agricultural interest served by the fair . . . a complex of interests . . . educational, technological, and economic." As the state became more industrialized and agriculture became more mechanized, industry began to constitute a larger part of the fair. In addition, as cities grew, fairs needed to attract those whose primary interest lay in industry, commerce, and the arts. Neely concurs, saying, "If the state fair was to profit by their presence, it was compelled, therefore, to cater to urban interests." The Missouri State Fair's exhibits over the years have reflected the changes in Missouri's products, while maintaining the importance of agriculture to the state's economy.

The Way We Used to Do It

At the 1976 fair, reporter Ron Jennings interviewed visitors who described the exhibit of antique farm implements as "one of the best things ever to happen at the Fair." Jennings also recorded Sam Templeton of Green Ridge explaining the machinery to his grandson: "You see, son, all of this here is what you've read about in school books come alive . . . You've heard me talk about such things, I mean working in the fields like we used to. Now you really know what I mean. Leastways, I hope you do. I hope you remember it, too." Templeton identified a "burnt-reddish, rust speckled implement" as a horse drawn cultivator: "Three discs on a side that could be set off at varying angles like, uh, this and you could also control the elevation of the discs by yanking the control stick this way. Understand now, that's the way we all used to have to do it when I was your age. Why me and the others, we'd work all day behind one of these and not think a thing about it."

Two very tired judges rest.
(M. S. A.)

Below: Pioneer Feed exhibited its products in a tent.
(M. S. A.)

Agriculture

The years between 1900 and 1917 were, according to *University of Missouri Bulletin 701*, a period of "moderate farm prosperity." Land prices were increasing, doubling between 1914 and 1920, the railroads were providing more lines, and farmers were becoming more aware of "conditions which gave the highest return for land and labor." However, the percentage of Missourians living on farms declined from 63.7 percent in 1900 to 53.4 percent in 1920. In spite of the decline in farm population, production of corn, oats, wheat, and hay increased steadily as farmers learned better methods. The fair enhanced production methods by demonstrating better methods and by establishing criteria by which crops could be "graded." Soybeans, introduced to Missouri in 1909, were first exhibited at the 1913 fair. At the 1915 fair, University of Missouri professors lectured on the use of fertilizers, testing soil acidity, silage crops, selection of seed corn, and treating seeds, thus fulfilling the fair's purpose as an educational institution. Clover and lespedeza highlighted the 1922 University exhibit.

Livestock production was becoming more scientific, too, as methods to prevent disease were developed. Breeders showed pure bred livestock, including the popular breeds of beef cattle—Shorthorns, Herefords, Aberdeen Angus, and Galloway; dairy cattle—Jerseys and Holsteins; hogs—Poland China, Duroc Jerseys, Berkshires, Chester Whites, Hampshires, and Yorkshires; and sheep—Shropshires, Hampshires, Oxfords, Cotswold, Dorsets, Southdowns, Rambouillet, and Delaine Merino. Missourian Augustus Busch, who owned one of two herds in the nation, showed Dexter cattle, a small "black and sleek" breed in 1915, but Dexters did not become popular with Missouri stock owners because of their size. Poultry raisers exhibited 102 varieties of chicken, six varieties of turkeys,

You Will Learn a Lot at the
Power Farming Show

How do you know whether a tractor would be impractical on your farm? What makes you think it would lose money for you? Just because your neighbor does not use a tractor for his farm work doesn't prove that you cannot either.

The only way to find out whether you can use a tractor profitably is to study its work, watch its demonstrations in the field and with belt and power work, then draw your own conclusions as to its value to you.

Have you ever done this? Now is the time to "be shown" one way or the other. In the Power Farming Show at the Centennial-Fair there will be scores of different types of tractors at work all the time. They will give practical plowing, farm-power and belt demonstrations. You can come and see for yourself the adaptability of a tractor to your farm work.

Reduced Railroad Rates

MISSOURI CENTENNIAL
EXPOSITION and STATE FAIR
Sedalia - August 8-20

MISSOURI CENTENNIAL EXPOSITION *and* STATE FAIR

The fair presented instruction in new farming methods during the early twentieth century. (M. S. F.)

This Duroc sow was champion of the world. (P. C. H. S.)

DOTY, 37472,
Grand Champion Duroc Sow of the World, 1904.
OWNED BY
McFARLAND BROTHERS,
SEDALIA, MISSOURI.

eleven varieties of ducks, seven varieties of geese, thirty-two varieties of pigeons, as well as pheasants, guineas, and peacocks. Two types of ferrets and four breeds of rabbits were on display. Livestock breeders saw a dip tank enabling them to protect their stock from Texas fever; in 1905, poultry raisers learned about "Chicken Raiser's Success," a medication guaranteed to "save them from disease." University lecturers provided less commercial advice, speaking in 1915 on balancing feed rations, feeding for egg production, recognizing and eliminating unprofitable cows, poultry diseases, hog cholera, feeding silage, and dairy rations.

Machinery was still primarily horse drawn, but was more efficient. Implement dealers exhibited stump pullers, "Barley's 20-Foot Giant Hay Stacker," and "Bowsher's Combination Feed Mills, sold with or without elevator." The 1907 fair boasted the first display of a "real milking machine in operation"; the milking parlor's display still interests city dwellers. In 1915, seventy-eight machinery manufacturers exhibited at the fair. The Altman-Taylor Company of Mansfield, Ohio, showed steam traction engines for threshers and plowing, portable saw mills, and gas tractors. By 1917, tractors had been developed, and the fair boasted that its "biggest and best farm implement show in the West" featured "The TRACTOR, the SILO (the farmer's best friends), together with other modern appliances for farm operation." J. Kelly Wright conducted a series of lectures on silo construction and use in a specially designated tent. The Fordson Tractor and a "Made in Missouri Machinery Show" highlighted the 1918 fair's exhibit.

During the 1920s, the popularity of automobiles, especially the Ford Model T, had reduced the isolation of many farms, as did the telephone. Improved roads followed the increased use of automobiles. Rural population continued to

Top: Rock Island Plow Works
distributed advertising cards at
the fair. (V. B.)

Above: The American Harrow
Company advertised in the 1907
Premium List. (M. S. F.)

Left: The Big 4 Tractor, advertised
at the fair, was not practical and
was not widely used. (M. S. F.)

decline; by 1930, less than half of Missourians lived on farms. The number of tractors increased, according to *Bulletin 701*, from 7,200 in 1919 to 25,000 in 1930, but did not completely replace horses and mules in farming. Although the number of tractors increased dramatically, Kirkendall points out that fewer than ten percent of Missouri farms had tractors in 1930. During the 1920s and 1930s, hybrid corn was introduced and became widely used. All these events affected the fair. The mules show remained popular, as did the draft horse shows and pulling contests. B. H. Heide, manager of the International Live Stock Exposition in Chicago, praised the 1922 exhibits as "superior." Professor H. H. Krusekoff charted soil fertility and corn diseases, causes of which had only recently been discovered. The University Building hosted a demonstration of "practices that pay" outlining techniques to make "the home, the farm, the garden, the field and herd and flock" more profitable. Automobile makers displayed their vehicles in an "automobile beauty pageant." The 1930 fair included a parade of farm machinery through the grounds; other vendors displayed seed, feed, feed supplements, wire stretchers, storage batteries, and oil burners.

The economic decline of the 1930s may have reduced the number of exhibits at the fair, but the exhibits maintained high quality. The six acre machinery exhibit in 1933 was "one of the largest displays of new and modern machinery seen at a Missouri State Fair in past five years," noted the *Sedalia Democrat*, which also praised the mule show. One million dollars worth of livestock—"some of the finest in the country"—paraded in front of the grandstand. In 1937, films on agricultural subjects were shown each day at the Veterinary Building.

Ironically, the pressures on agriculture created by the Depression and World War II aided Missouri agriculture, as scientific methods of increasing production while saving labor and costs were developed. The percentage of farms with tractors increased, according to Kirkendahl, from 16.4 in 1940 to 27.8 in 1945. *Bulletin 701* notes the steady increase in tractor use freed for crop production many acres of land formerly used to raise grain and hay for mules and horses. The G.I. Bill enabled many young men to attend college, and enrollment in agriculture programs increased. These newly educated farmers understood scientific breeding and crop production, the necessity of controlling soil erosion through terraces and contour planting, and the importance of

Top: These sheep were shown during the 1930s. (M. S. A.)

The steam-powered tractor would eventually be replaced by gasoline-powered tractors that were more efficient. (M. S. F.)

A judge examines a chicken in 1939. (M. S. A.)

Below: Judges and spectators examine a group of mules. (M. S. A.)

meeting market demands for their products. The fair performed an important educational function by demonstrating conservation techniques, newly introduced breeds of livestock, and higher quality varieties of grains to the state's farmers, especially those outside the university.

Quality and quantity of product determined what would be produced, and what new products would be introduced. By 1947, the swine exhibit had expanded to included Spotted and Hereford hogs. Dairy exhibits had added Guernsey, Brown Swiss, and Milking Shorthorns. Polled Shorthorns and Red Polled cattle expanded the beef cattle department; sheep exhibits now included

Corriedale and Columbia sheep. A dairy goat department featured Nubian, Toggenburg, and French Alpine goats. The *Premium List* advertised sales of breeding stock and agricultural journals. The 1945 American Standard of Perfection established criteria for fowl judged at the fair. Machinery exhibits increased as well. The 1952 machinery exhibit included fifty-three exhibitors with seventy-five carloads of machinery, including several combines.

Throughout the 1950s and 1960s, the livestock classes continued to expand. In 1956, Suffolk sheep were added and the rabbit show increased to nine breeds.

Top left: Showing both a live chicken and a dressed chicken allowed judges to evaluate the meat's quality. (M. S. A.)

Above: A young man displays his Charolais cow. (M. S. A.)

Left: A Corriedale sheep is on display. (M. S. A.)

Below: The Santa Gertrudis is a large, heavy breed.

During the 1950s, poultry breeders at the fair extolled the "chicken of tomorrow," able to reach broiler size of two pounds more rapidly through the use of higher quality feed. By 1962, twelve breeds of rabbits, seventeen breeds of ducks, and eleven breeds of geese were recognized. Swine added Yorkshires and Landrace hogs, longer, leaner hogs with high litter size. Charolais cattle, imported from France after World War II, were part of the livestock exhibit in 1967. As farm size increased and the pool of available farm laborers decreased, farm machinery became larger and more elaborate. Farm implement displays expanded to include larger and more versatile tractors, combines, planters, and plows.

The 1970s through the 1990s revealed changes in livestock production. The market demand for larger, more heavily muscled animals that produce more meat, and animals tolerant of heat and insects spurred the interest in "exotics" (breeds originally developed in other countries) and cross breeds. Santa Gertrudis cattle, a cross between Brahman and Shorthorn; Limousin, introduced from France in 1971; and Simmental, cattle originally from Switzerland, showed at the 1975 fair. Brangus, a cross between Brahman and Angus; Chianina, introduced from Italy in 1971; and Maine Anjou from France were added to the livestock show in 1982. By 1987, Brahman from India, Gelbvieh from Germany, Longhorn, Salers from France, and Simbrah, a cross between Brahman and Simmental, had been added. Swine also reflected the market demand. A leaner hog that would produce less lard and more lean meat was demanded by a more health conscious public. Chickens and egg production became less a sideline managed by farm women and more an agricultural industry, though poultry raising continued to be a popular hobby. In 1996, the poultry department recognized 401 varieties of chicken, eighteen of duck, and seven of turkey. The Goat Department added Angora, Kinder, and Boer breeds. The Beef Cattle Department dropped Brahman and Longhorn classes, but added Belgian Blue Cattle.

Fruit, especially apples, peaches, and grapes, form an important part of Missouri's horticulture and are on display in the Agriculture Building, as are vegetables from Missouri's gardens. The number of varieties of apples and peaches has declined, but the quality and size have increased

This Berkshire hog demonstrates the lean, long body expected now. (M. S. A.)

through the years. Vegetable exhibits have changed as market demand has changed. For example, at early fairs, parsnips were displayed and zucchini were not; parsnips have been dropped but zucchini added. Prizes are still given for the largest watermelon and the largest pumpkin, acknowledging the importance of size as a criterion for judging some vegetables. The Floriculture exhibit has now expanded to include several separate shows co-sponsored by various associations. Roses, gladiolus, African violets, orchids, daylilies and lilies, potted

plants, and dahlias are on display in separate shows. Popular garden flowers such as asters, celosia, cosmos, marigolds, petunias, snapdragons, tithonias, tuberoses, and zinnias show as specimen plants. Ornamental trees and flowering shrubs also have a place in the show. The Federated Garden Clubs of Missouri demonstrate flower arranging, propagation, and care.

Other farms and other critters

One of the most popular exhibits at the 1941 and 1942 fairs was "Uncle Ezra's Farm," a farm scene complete with "electrically-operated, hand-carved animals, people and equipment." The six foot by nine foot farm valued at $10,000 featured a tractor, a pond of fish, a man fishing, piglets and sow, a cow swishing its tail, ducks, and geese. In addition, imported animals such as llamas, camels, and emus and native animals like bison and elk appear at the fair as their proponents try to encourage a more diversified agriculture.

In 1923, the State Fair Dog Show opened. Over two hundred of the "finest dogs in Missouri" paraded through the show ring. Judge Charles D. Hopton of New York City was impressed with the "exceptionally good" quality of Missouri's dogs, which would "favorably compete in their classes in the largest shows in the east." A dog show of a different sort came to the fair when raccoon hunters staged races; the dogs raced through a pool following a caged raccoon to see which dog was fastest. In 1976, 260 dogs competed.

Wild animals also visit the fair. In 1923, Governor Arthur Hyde christened a baby lion belonging to Gentry Brothers and Patterson's Famous Shows with the name "Sedalia." In the 1990s, tigers paced their cages set up west of the Agriculture Building, enticed by the scent of poultry displayed in the Poultry

Top, left: The Poultry Building hosts many breeds of chicken. (M. S. A.)

Top, right: Sharon Meahler cheers her chicken's performance. (T. P.)

Combines and corn pickers highlight the machinery exhibit in 1968. (M. S. A.)

Missouri Farmers Association

During the 1870s and 1880s, Missouri farmers had joined with farmers in other states to form the Patrons of Husbandry, more commonly known as the Grange. By the early twentieth century, the Grange had become less effective in its ability to influence government policies that affected farmers. Around 1914, William Hirth began to organize what would become the Missouri Farmers Association, whose purpose, Kirkendall quotes, was to "assure to the agricultural interests of Missouri the reward and recognition to which their fundamental importance entitles them" and to be sure farmers earned "reasonable" profits from their labor.

By 1923, MFA had 42,000 members and managed cooperative livestock shipping facilities, grain elevators, and farm supply stores. During the 1940s, membership grew, and by the mid-1950s, MFA boasted 150,000 members. The MFA met for its annual convention on the fairgrounds, with its meetings often beginning on the last day of the fair and continuing for one to two days beyond the end of the fair. The group became active politically, supporting policies and candidates it believed would benefit farmers.

Despite the difficulties faced by farmers since the farm crisis years of the 1980s, the MFA continues to be a significant factor in Missouri agriculture.

MFA distributes balloons, crowns, and fans at the fair. (M. S. A.)

A gardener could learn how to raise gladiolus. (M. S. A.)

Right: The Floriculture Building is bedecked with vines in this 1950s photo. (M. S. A.)

Building to the south. Alligators, snakes, sharks, and dolphins, and the people who handle them, have amazed visitors since early fairs.

Commerce

In 1927, the *Sedalia Democrat* editorialized," The state fair was instituted and has been maintained largely in the interest of the agricultural industry of Missouri . . . But the fair has become increasingly representative of the activities of all the people of the state. Its interest is for the city man as well as the farmer. It relates to progress in manufacturing, in education, in art and still other lines of endeavor." Commercial exhibits show newly developed products designed to make life easier. For example, the Delco Batteries advertised at the early fairs provided a way to store electricity produced by a generator so as to provide electricity to homes beyond the reach of power lines. A "Carbide Cooking and Heating Plant" shown in 1918 demonstrated one method of providing light and heat. In 1935, the Economy Gas Generator Company of St. Louis showed its "individual gas making plant for farm homes" that burned wood to provide gas for refrigeration, heating, lighting, and cooking. The Rural Electrification Administration enabled the percentage of Missouri farms with electricity to increase from 17.7 percent in 1940 to 31.5 percent in 1945 to 70 percent in 1950. Appliance manufacturers displayed the newest household appliances, and home economists demonstrated practical uses of the new technologies.

Other Missouri industries promoted their products. The 1917 Premium List advanced Missouri's lumber and mining industries by requesting counties and companies to make "elaborate exhibit of various timbers . . . of the quality and variety of building stone . . . and the valuable minerals that may be mind with profit." 1918 vendors sold stoves, knives, jewelry, microscopes, and even pianos. The 1930 *Program Book* printed a list of commercial vendors, including

Leather goods are for sale at the fair. (M. S. A.)

Dogs test their ability to swim in pursuit of a raccoon. (M. S. A.)

Kenmore advertised its very modern kitchen in the Rural Electric Cooperatives Building. (R. E. C.)

Top left: Instant Heat provided hot water for the home. (M. S. A.)

Above, right: The *Premium List* in 1918 emphasized women's accomplishments. (M. S. F.)

Two men demonstrate cooking in the Home Economics Building. (M. S. A.)

a beauty shop in the Varied Industries Building, a barbershop on the grounds, the Dorn-Cloney Laundry, and sellers of wire jewelry, one-minute pictures, a braider attachment for a sewing machine, hair curlers, and stoves. Osteopathic medicine advertised itself. Fairgoers could have their psyches examined and their palms read by Dr. F. H. Ruble of Kansas City, who offered his services as a psychologist and palmist. In 1936, just three years after the repeal of prohibition, the Missouri Brewers Association mounted an exhibit of oil paintings illustrating the six thousand year history of brewing. The Missouri wine industry, slower to recover from prohibition, advertises its products at the fair. Flour mills, leather crafters, encyclopedia publishers, sewing machine makers, and furnace and air conditioner companies still try to entice visitors to stop, look, and buy.

Homemade Quality

In 1916, Sedalia's ladies' newspaper, the *Social Messenger*, praised the "array of exquisite fancy work, embroidery, crocheting, and needlework...most beautifully arranged." Needlework remained both a practical and decorative skill; in the 1940s the Staley Milling Company gave special prizes to clothing, embroidery, quilts, and rugs made from flour sacks. Sugar, flour, baking powder, and cocoa companies, meat processors, pasta manufacturers, packaged pie crust makers, honey growers, and a host of other food processors encourage the use of their products through specialized cooking contests. Bar-B-Que contests bring out the secret sauces and grilling techniques of the best backyard cooks. Displays in the Home Economics Building demonstrate the skills of Missouri quilters, knitters, seamstresses, cake decorators, bakers, and cooks.

The Arts

The Fine Arts exhibits have revealed the changes in style as art has moved from realistic to impressionistic to abstract. Both amateurs and professionals have shown works since early fairs, but the nature of the shows has changed. At early fairs, paintings and drawings were displayed with pyrography, leatherwork, and decorative arts. Later, the Womans Building was designated as the place for decorative arts and crafts, and the Fine Arts Building housed drawings, paintings, and china painting. Decorative arts are now displayed in the Home Economics Building. In 1991, with the help of the Missouri Arts Council, the Fine Arts Department changed its format. The Missouri 50 Show, held upstairs in the Fine Arts Building, features juried works by fifty of Missouri's premier artists. The ground floor of the building hosts the professional and amateur shows and the China Painting Department.

Several well-known Missouri artists have exhibited at Missouri State Fair. In 1933, Wilbur Phillips won acclaim with his paintings *Saturday* and *Ozark Farmer*, which showed the "plain reality" of the "struggle for existence" in southern Missouri. Trew Hocker won in 1935 with his painting *Street Singer*. Darrel Crider's installations of wooden stick figures and distorted chairs have amused viewers, and Elizabeth Gray's brightly-colored collages remind viewers that senior citizens have incredible talent.

Perhaps nothing at the Fine Art Show has equaled the events of 1939, when a

These paintings were exhibited at an early fair. (M. S. A.)

Inmates from the Missouri State Penitentiary won ribbons for their paintings. (M. S. A.)

A potter demonstrates throwing pots while working as an artist in residence. (M. S. A.)

primitive painting of a farm by Flora Lewis won first prize. Lewis, wife of a Marshall veterinarian, had no training as an artist, but had exhibited work at other fairs, and had won a blue ribbon for a painted pillow entitled "Suffer the Little Children to Come Unto Me" at the Chicago World's Fair in 1935. Though Lewis often painted highly symbolic pieces for use in area churches, *Farm Scene* was realistic. The art students and professional artists were highly indignant, and the press criticized the work and the decision of judge Austin Farley, professor at Stephens College, who called the painting "the finest example of primitive art I ever have seen." Police had to control the crowds who came to see the painting. Mrs. Lewis answered her critics by simply saying, "They judged it, didn't they." In 1952, Dorothy Truitt, who had been superintendent of the Art Department in 1939, commented that "she was just as shocked about the picture and the judging as everybody else," but after having lived for many years in California, she had learned to appreciate the importance of primitive art.

Missouri Farm Bureau

Join with others — Have your say — Guard your freedom — All the way.
Farm Bureau

In the 1950s, these words, painted on a series of signs, graced Highway 65 in southern Missouri. They represented the Missouri Farm Bureau Federation, a farmers' organization that had developed during the early twentieth century. During the 1920s, Farm Bureau boasted, according to Kirkendall, forty-one county units and fifty thousand members. It was part of the American Farm Bureau Federation, and lobbied for better roads and schools, and fair taxes for farmers. It also, Kirkendall notes, lobbied on behalf of fairs that would enable

farmers to learn about better agriculture by showing their products. The Farm Bureau actively encouraged the University Extension Service, both with support and monetary contributions.

During the 1940s, the Farm Bureau increased its membership and its power, still exerting influence on government policies that affected farmers. As farming came to be more affected by government policies, the group continued to grow, and at the end of 2001, had 94,766 Missouri families as members.

Music has been a part of the fair since its beginning. In the early years, an official band was chosen to perform concerts on the grounds at designated times during the day. Vocal concerts drew large crowds. In 1916, the *Social Messenger* praised Thiviu, the director of music's "fine band, artistic singers, and charming personality . . . may he continue to furnish the music of our fair." Thiviu brought an orchestra, several opera singers, and a ballet company to perform each year, and often invited the Sedalia Ladies Musical Club to sing with his ensemble. High school bands were invited to parade through the grounds and to perform concerts under the leadership of University of Missouri music professors.

In 1924, the State Fair Music Contest, co-sponsored by the Missouri Federation of Music Clubs, began. The *Sedalia Democrat* acknowledged that such competition would "give the young musicians in Missouri something to work for and compete for and keep enthusiasm at a high pitch." Young people performed piano solos, duets, vocal selections, and violin numbers. In 1925, "Missouri ranked first in attention given to music last year," and it would continue to "lead the state fairs again in this important feature of the exposition." The Music Contest, originally held in the Womans Building, is now held at State Fair Community College. The Old Fiddler's Contest provides a different type of music. Children and adults play the lively hoedowns, waltzes, and other tunes, accompanied by guitar, banjo, dulcimer, mandolin, or other acoustic instruments.

Flora Lewis' *Farm Scene* was an excellent example of primitive art. (M. S. A.)

Percy and Flora Lewis enjoy a hot dog after her victory in the art contest. (M. S. A.)

Health Exhibits and Healthy Babies

Health care organizations have exhibited since the first fair featured an x-ray demonstration. Patent medicine hawkers advertised cures "good for man or beast." The Department of Health provided exhibits of Pure Foods contrasted

These young people won in the Music Contest for their violin solos. (M. S. A.)

Third-prize babies won drinking cups. (V. B.)

with adulterated and unsafe foods, an important concern after the passage of the Pure Food and Drug Act. In 1937, the State Board of Health began inspecting eating places, restrooms, and toilets for sanitation. The Missouri Medical Association offered a display of pictures illustrating the stages of alcoholism. The American Cancer Society gave information. In 1941, when polio was feared and vaccines had not yet been developed, the National Polio Foundation had a booth outlining its work. The 1951 fair set aside half of the Education Building for a "Hall of Health" with exhibits by Missouri hospitals, nursing schools, the Tuberculosis Association, the State Dental Association, and a "most interesting display" by the American Medical Association of "medical fakes through history." Most of the phony curative devices shown in the exhibit were called "fugitives from a penny arcade," by medical histo-

Farm Families

For many years, the Missouri Agricultural Extension Service has recognized an outstanding farm family from each county. The families, chosen by county extension councils, exemplify Missouri agriculture. They are actively engaged in farming and rely on the University Extension Service for advice and technical assistance, and support the University Extension activities. Their children are active in 4-H or FFA. In addition, the farm families are active in local affairs, involved in school, church, and charitable organizations in their communities. Recognized as leaders, farm family members are respected by their neighbors.

The families come to the Missouri State Fair, receive recognition plaques, and enjoy a dinner compliments of the Missouri Farm Bureau and the Missouri State Fair.

This Farm Family was honored at the 1966 state fair. (M. S. A.)

A healthy child receives an award in 1947. (M. S. A.)

rian Oliver Field. In 1952, 1,332 people received chest x-rays in a mobile unit on the fairgrounds, and an electrocardiograph display warned fairgoers of the early warning signs of heart disease. In 1985, cholesterol screenings allowed fairgoers to assess their diets and health before proceeding to the corn dog stands.

Children participated in the health exhibits by becoming exhibits. The Board of Directors considered having a Baby Health Contest prior to 1920, and by 1921, the contest was firmly in place. The Department of Health offered free health examinations for children, who were judged and awarded prizes. The blue-ribbon babies were examined for "nutrition, dental condition, nose and throat, mental development, immunization against diphtheria and smallpox, and regular practice of health habits." Nine-point children were evaluated on their "nutrition, posture, vision, hearing, teeth, throat, diphtheria immunization and smallpox vaccination." The governor presented prize-winning children with ribbons and trophies. The concept of eugenics—the scientific breeding of better humans—lent ominous overtones to the health contests, but most parents simply saw them as a way to get a free examination for their children and a way to show off their beautiful babies.

The 1937 Healthy Babies with their trophies. (M. S. A.)

Below: A proud mother holds her baby who won the 1957 baby contest. (M. S. A.)

Hobbies

Through the years, various hobbies have been displayed at the fair. During the 1930s through 1960s, stamp collectors hosted the Missouri State Philatelic Show. The 1941 show displayed "several thousands of dollars" worth of stamps, and boasted an increase of 115 frames over those shown in the 1940 show. The 1951 show

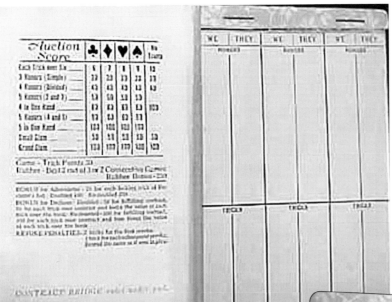

A Gin Rummy Tournament was part of the fair in the 1920s. (V. B.)

included thirty-two different classes. Antique dolls, needlework, toys, and furniture have been favorite exhibits for collectors. Checker players participated in tournaments during the 1936 fair, and card players in gin rummy tournaments in 1920s. Horseshoe pitching contests, long a part of the fair, now offer $2,070 in prizes and are sanctioned by the National Horseshoe Pitchers Association.

It is difficult to imagine something that has not been exhibited or sold at the Missouri State Fair. The fair has truly "extended . . . its sphere of patronage," as the 1905 *Premium List* said it must, and offers something for everyone.

CHAPTER 8

"IN ITS ABILITY TO TAKE ON EDUCATION":

Horse Shows and Races, Rodeos and Mules

FAIR HISTORIAN WAYNE CAMPBELL NEELY POINTS OUT THAT BY the late nineteenth century, the racetracks had become a "generally accepted, if not an essential, part of the fair." However, racing had to be justified as a part of the educational aspect of the fair, as a way of allowing visitors to see the improvements in the animals through appropriate breeding and training. The "development of speed and stamina in light-harness horses . . . and the development of size and strength in draft horses," Neely suggests, constituted important factors at a time when horses still provided most transportation and power. In 1900, Missouri was known throughout the nation for its saddle horses and mules, with 248,850 mules and over 750,000 horses on Missouri farms, according to historians Lawrence Christensen and Gary Kremer. The

Above: Saboteur had developed the "speed and stamina" necessary to win this race. (M. S. A.)

Facing page: The foal this boy is patting would become both a friend and coworker. (M. S. A.)

automobile and tractor eventually would replace the horse and mule, but by 1950, historian Richard Kirkendall points out that one-third of Missouri farmers still did not own automobiles.

The importance of horse racing to the Missouri State Fair is readily seen in the time and money spent on the racetrack prior to the first fair. The track was the first element of the grounds built, and $20,000 was spent on its construction. Barns for show horses and mules soon followed, further emphasizing the role of horses and mules in Missouri and at the fair.

The state fair was to be funded by Breeder's Fund, monies accrued from licensing racing and betting throughout the state. Betting on horse races had raised moral issues since before the establishment of the fair, but fair supporters were able to deflect such questions by pointing out that betting was legal. In 1903, the fair banned gambling on its races, noting "That good contests of speed without gambling can be had was demonstrated, as no pool selling or gambling of any kind was allowed." The issues of the Breeder's Fund and the morality of betting came to a close in 1907 when state statutes prohibiting betting on horse races were enacted. The Union Park Race Track in St. Louis was sold to the Louisiana State Fair, and the Missouri State Fairgrounds became the premier track in Missouri.

Races, planned each day of the first fair, were held as often as weather allowed. The races included trots, paces, and runs. American Standardbred horses participate in trots and paces; thoroughbreds in runs. A trot is a one-mile race between trotters, horses who move their legs diagonally, so that the right front leg and left hind leg strike the ground simultaneously, and then the

Draft horses test their ability to pull heavy loads. (M. S. A.)

Below: This 1908 trade card shows the racetrack. (V. B.)

left front leg and the right hind leg hit the ground. A pace is a one-mile race between pacers, horses who move their legs laterally, so that the right front and right hind leg hit the ground together, then the left front leg and the left hind leg strike simultaneously. Trotters and pacers race pulling two-wheeled carts called sulkies. Runs are races of varying lengths between horses with mounted jockeys. Well-known horseman Lee Clement, editor of the *Kentucky Stock Farmer*, praised the state fair track as "one of the fastest and best tracks of its kind in the country." A fast track, smooth, with no puddles, excessive dust, or uneven patches, allows the horses to perform at their maximum potential. In 1901, Clement suggested that more money be appropriated for prize money in order to "secure the stakes offered by the *Horse Review* and the *Kentucky Stock Farmer* and other associations." Higher purses would also draw the "star performers" of the racing world, so that those responsible for "developing Missouri colts" would be "schooled...to know and appreciate what education is and the strength it gives to them in educating the highest class of our animal friends and co-workers." He advocated a circuit of races at state fairs throughout the Midwest, so that visitors from other states could see what Missouri was capable of producing. "We can yet improve the horse," Clement insisted," in his speed and in his susceptibility to take on education."

Over the years, Missouri horses would show what they had learned. Clement hoped breeders and trainers could produce a trotter capable of racing one mile in two minutes. Noting that Surpol, "a 2:10 trotting stallion" lived in Missouri, he urged "liberal premiums to bring these sires, raced and retired to

stud" and their offspring to the Missouri State Fair. By 1903, the purses offered included $1,000 for the 2:30 trot, and $500 each for two trots and a free-for-all pace. Running races, not as popular as harness races, drew $200 purses. Entry fees and the large numbers of horses participating funded the prizes, and the *Sedalia Democrat* was happy to report that "the races paid for themselves."

Racing remained an essential part of the state fair. In 1911, the *Sedalia Democrat* announced, "the speedway this year will climax all previous years, as the princely purses have attracted the kings and queens of the American turf—a profusion of the greatest races ever seen here." As automobiles became more widely used, the necessity of developing the horse's speed and stamina for work came to be less important, but love of horses enabled racing to continue. In 1913, the fair advertised the "best horse racing in the country." In 1916, the editor of the *Social Messenger* admitted she was "no judge of a race," but said the races that year were "better than average." 1918 marked "the beginning of a new era in horse racing." Prizes had increased to a total of $16,800 for the week, and had "brought new owners with new horses to Missouri's mile track." Automobile races on Saturday had been eliminated, and the money put toward increasing the horse racing program to six days.

Dr. J. P. Thatcher advertised his championship horses in the 1907 *Premium List*. (M. S. A.)

Winning horses are recognized with a bouquet and hefty purses. (M. S. A.)

During the 1920s, purses and crowds became larger. The 1922 fair featured four afternoons of racing with a total of $10,000 in prize money. "Approximately 100 running horses are on the grounds and are being conditioned for the opening day races. A number of harness horses are also in the barns and will be put through their limber-up exercises preparatory to the opening of the harness races on Tuesday of next week," wrote the *Sedalia Democrat* in 1925. Newsreels shown at theaters included track results; in 1927,

Governor Baker's daughter "presented a huge floral horseshoe to the winner as movie cameras cranked." Crowds began to gather around noon and by race time, both the amphitheater and bleachers were full; over eight thousand people saw the Missouri Cup Trot, as well as other trots and paces, and two running races.

Favonian paces his way to victory. (M. S. A.)

Below: Trotters pull sulkies as they race. (M. S. A.)

Although fair attendance dropped during the 1930s, races continued to draw large groups of participants and spectators. In 1933, "Outstanding horse races brought horsemen of the country to Sedalia and the fair. They mingled with the Missouri people, some of whom did not know horseflesh so well, but did realize they were fine horses and they enjoyed seeing them." Clement's dream of a Missouri State Fair race breaking pacing or trotting records still had not been reached, but in 1935, a horse named Jerry Patch set a record with 2:07 1/2 in the second heat of a two-year-old pace. Horse races often alternated with vaudeville acts in front of the grandstand. The 1936 fair featured a "running horse day, with a full afternoon program of fast running events on the new half-mile track plus band music and six stellar stage acts."

During the 1930s, Works Progress Administration workers built a half-mile track, and in 1947 a Litzenberger starting gate, the first of its kind in the Middle West, was installed. The creation of the Grand Circuit of Harness Races increased the level of competition at the fair. In 1951, four out of the five horses running in the three-year-old trot had won the Hambletonian, the premier harness race in the U. S. In 1952, the fair recognized the work of Ed

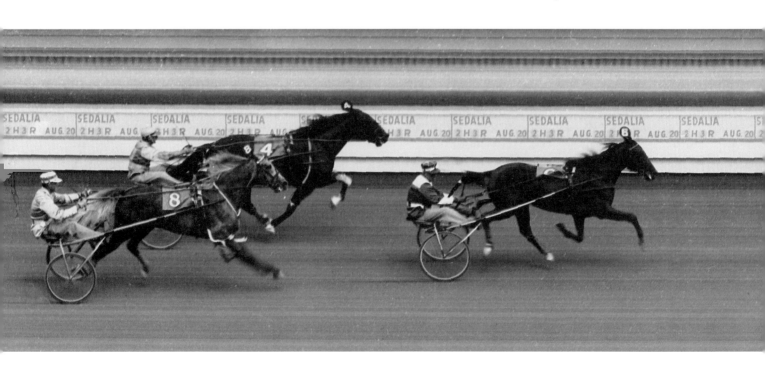

Duensing, of Jefferson City, long-time superintendent of speed department, for his thirty years service as a fair official. Although the fair's races were no longer a part of the Grand Circuit after 1963, harness racing remained part of the fair, and in 1964, four days were devoted to racing. In 1976, the racing program included three days, a total of twenty-four races, of miniature harness horse racing, as well as twelve harness races with purses ranging from $2,000 to $1,000. The All Breed Missouri Horse Racing Association sponsored running races with Thoroughbreds, Appaloosa horses, Paint horses, and Quarter horses. The legalization on pari-mutuel betting at the state fair in the 1980s brought more people to the track. On-track betting was discontinued after only a few years. Racing has diminished somewhat at the fairgrounds, and several of the original horse barns have been torn down. Two of the 1901 barns remain, and racing continues at the fair, with two days and fifteen races devoted to the sport.

Horses line up behind the starting car. (M. S. A.)

Women generally did not participate in horse racing. Sedalia had seen races between women in its local fairs held in Association Park before the establishment of the state fair. Races between "equestriennes" riding sidesaddle were disapproved of by those who felt the ladies revealed more of themselves than was necessary. However, by 1937, female jockeys rode astride in jodhpurs. Lillian Jenkinson rode her father's horses in four races, placing first in two and second in two. Women's participation in racing remained unusual, but women's involvement in horse shows had a long-standing history.

Horse Shows

Early fairs recognized the importance of horses for labor and transportation, and offered competitions for draft horses and light horses—saddle horses and carriage horses called roadsters. In the light horse classes, roadsters "shown in harness" were judged for their "speed, style, form, and action." Roadsters and carriage horses "with appropriate vehicles" were judged by the same criteria. French and German coach horses, shown "to bridle," demonstrated their "form, style, and action." The classifications used in early fairs reflect the uses of horses at that time, and emphasize both speed and beauty.

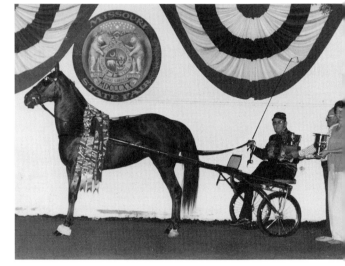

Top: Erwin and horse Goodtime receive a tray as part of the prize. (M. S. A.)

Harness horses show in the Coliseum. (M. S. A.)

Top left: Sedalian Frank Erwin had a long career as a harness racer. (P. C. H. S.)

Above: Governor Hearnes presents a trophy to Wichita Witch. (M. S. A.)

By 1911, the horse show entries had increased so much that a third show had to be added and barn space was outgrown as "noted horsemen and horsewomen of the country entered their famous horses" in shows that "promise to be big features of the fair." By 1915, horse shows "brought out some of the real classy show horses." The eleven classes in the Wednesday night horse show included harness horses, harness horses driven by women, three-gaited saddle horses ridden by women, Shetland pony pairs in harness, roadster stallions in harness, and a junior saddle horse stake with a prize of $135.

Exotic dancer Sally Rand presents a trophy to a winner in the 1950s. (M. S. A.)

The horse receives a garland; the driver receives a ribbon and a trophy. (M. S. A.)

The night horse shows proved popular, as they enabled working people to attend and allowed all to take advantage of the cooler evening weather. Those who attended a horse show generally saw more than the horses, as band or grand opera music and vaudeville acts were presented between the individual classes, and fireworks displays ended the evening. In 1921, the number of classes of horses had expanded, and included a category of "carriage horses, pairs" which "must have long tails." In addition, the prize money increased. Total premiums for light horses in 1922 totaled $500, with $2,500 allocated for draft and coach horses.

Many well-known Missouri horse breeders exhibited their animals at the fair. Tom Bass from Mexico, Missouri, considered by some to be one of the best trainers in Missouri, showed frequently. Bass was notable enough that the

Board of Directors recorded in their minutes of 1909 that he was bringing horses for a demonstration of their abilities. In a 1927 competition of five gaited saddle horses considered difficult to judge because of the quality of all the entrants, Bush McDonald, Tom Bass' black stallion, won first. Loula Long Combs, the "grand dame of Missouri horse-women" and owner of the famous Longview Farms in Lee's Summit, was honored in 1955 for her sixty year career as a breeder and shower of fine horses. Arthur Simmons of Mexico, Missouri, regularly showed at the state fair; in 1955, Simmons set a record by twice winning firsts in three shows.

Horsemen and women from outside Missouri showed at the fair also. S S. Lord of Fort Worth collected many ribbons—five firsts, one second, and one fourth— in the 1927 fair, to which he brought seventeen horses. The *Sedalia Democrat* praised the "175 head of the finest saddle horses" from fifty stables but lamented that limited space at the fairgrounds kept many horses away. The stables included Hall-Mar farms of Eldorado, Kansas; Mary Gwyn Fiers of Oklahoma City, Oklahoma; Twinwoods of Lincoln, Nebraska; W. M. Hynes Stables, the B. B. Tucker Stables, and the Three Bar Stables, all of Omaha, Nebraska; Mrs. R. W. Brooks of Colorado Springs, Colorado; W. N. Dalton of Kansas City, Missouri, and H. C. Bryant of Oklahoma City, Oklahoma. The three gaited saddle horse show was enhanced by the appearance of Roxie Highland, the champion three-gaited saddle mare of the United States. In 1935, horses from San Francisco, Boston, and two stables from Dallas participated; Mrs. W. P. Roth of San Francisco won first place in both classes of the national saddle horse futurity and set a record for winning first place in every futurity event and breeders' stake offered at the fair.

Horse shows changed over the years. Occasionally well-known horses or riders appeared. For example, in 1905, the Montana Girls showed their skill as riders of bucking horses. A palomino owned by department store

Breeder Art Simmons was active in Missouri horse shows for many years. (M. S. A.)

Below: The Tennessee Walking Horse is known for its distinctive gaits. (M. S. A.)

magnate J. C. Penney, originally from Hamilton, Missouri, performed a ten minute exhibition in 1941. Cilly Feindt demonstrated dressage, a type of riding emphasizing sideways movements, circles, and other intricate maneuvers. Although professional riders and trainers received most of the publicity for their appearances in horse shows, throughout the fair amateur riders and trainers, motivated by their love of horses, have participated in the shows. These dedicated individuals often surprised the professionals when they carried the blue ribbons out of the ring.

Shows also increased in size as more breeds became popular. In 1942, Walking Horses, a light horse breed noted for its distinctive gaits, the flat walk and the "extra-smooth, gliding" running walk, showed at the fair. In 1952, the Morgan Horse, an old American breed experiencing a renewed popularity, joined the horse show. In 1955, Saddle Club classes, including western horsemanship, men's western pleasure class, a matched western working pair, and reining horses class performing "intricate maneuvers" became a part of the horse show, further reflecting the use of horses for pleasure. The state fair horse show was designated in 1963 by the American Horse Show Association as one of fifteen "honor shows." Eight days were devoted to horse shows in 1964, with specific classes for walking horses, three gaited and five gaited saddle horses, harness and hackney ponies, hunters, jumpers, quarter horses and Appaloosas. Horse shows follow the rules of the American Horse Show Association and the various breed associations.

The year 1991 marked two important events in horse show history at the fair. Eighty miniature horses, under the auspices of the Heartland Miniature Horse Show, demonstrated their skills. The small horses, valued at as much as $10,000 each, were a hit with spectators and horse people alike. In addition, the Society Horse Show, noted for the formality and beauty of the horses, their trappings, and their riders, returned to the fair after an absence of several years. Doris Marks, superintendent, anticipated "some new and exciting saddlebred stables to exhibit this year." The show was dedicated to the Arthur Simmons family, stable owners from Mexico, Missouri. The Society Horse Show continues, this year honoring Paul McDannald, of

A girl practices for a show in the arena outside the Coliseum. (M. S. A.)

McDannald's Arabians in Sarcoxie, Missouri, for his thirty years' work with Arabian horses at the state fair. The Society Horse Show provides an opportunity for saddle-breds, foxtrotters, Morgan horses, Arabians, three and five gaited saddle horses, and saddlebred horses in harness. It also features a "country pleasure division" for horses "plain shod," with "full mane and tail" and riders dressed "conservatively" in "informal attire with coat and hat."

Mules, Jacks, and Jennets

In 1900, Missouri bred more mules than any other state in the nation. The sale of mules constituted an important part of Missouri's farm economy,

Above: Elaborate trappings are often part of a horse show. (M. S. A.)

Right: Women have partici-pated in the state fair horse shows since the beginning of the fair. (M. S. A.)

Top: A young rider receives a ribbon. (M. S. A.)

Above, left: Riders in the Society Horse Show are expected to follow traditional rules of dress. (M. S. A.)

Above: This winning team poses inside the Coliseum. (M. S. A.)

Left: Governor Hearnes admires a pair of Missouri mules. (M. S. A.)

A draft horse team performs in the grandstand in 1952. (M. S. A.)

Christensen and Kremer suggest, with Great Britain buying over 100,000 mules between 1899 and 1902 for use in the Boer War and the United States spending over $500,000 on Missouri mules and horses during the Spanish-American War from 1889 through 1900. In 1902, the fair's "splendid display . . . of mules and jacks" paid homage to the "animal which has done its share toward carrying the fame of the state abroad."

A mule is a hybrid animal, a cross between a male jack and a female horse. A female mule is called a jennet. The mule is more disease resistant than the horse, can tolerate heat and dry conditions better, is easier to feed, and remains useful longer. Missouri mule breeders paid close attention to the animals selected for breeding, and produced high-quality animals. The mule show, in which jacks, jennets, and mules compete, has remained a fixture of the state fair. In 1922, jacks, jennets, and mules gathered a total of $1,500 in premiums. Generally, mules, jacks, and jennets are shown at halter, but in 1947, a class was established for pairs of mules, weighing not more than 2,700 pounds "to wagon appropriately hitched." The "large number of four team mules" in the horse show in 1952 forced the judges to move from the Coliseum to the grandstand. When Harry S Truman visited the fair in 1955, he went into the stables to inspect the mules. Truman, who had grown up in Lamar, Missouri, understood the qualities of a fine mule and the importance of mules to Missouri agriculture. At the 1976 World Championship Mule contest, two thousand spectators gathered to see Missouri mules, and mules from other states as well, at their finest. Though mules are not generally used on farms now, the mule remains a symbol of Missouri tenacity—or stubbornness. The world's Championship Mule continues to be selected at the Missouri State Fair.

Draft Horses

Draft horses, called "heavy horses" in early fair *Premium Lists*, are large horses, averaging fourteen to nineteen hands high at the shoulders, bred for their size and strength. Developed in the Middle Ages as war horses, draft horses were improved by careful breeding during the early nineteenth century for use in agriculture and drayage. After being imported to the United States, they quickly became popular with farmers and teamsters. Draft horse owners who came to early fairs could see the best of the three most popular breeds of draft horses, learn about characteristics of quality stock, and test their animals'

Draft horses in show are evaluated on their appearance and ability to execute commands from their driver. (M. S. A.)

Top: Draft horses, originally bred for work, compete to see which horse can pull the most weight. (M. S. A.)

The driver rides a weighted sled in a draft horse pull. (M. S. A.)

attractiveness and confirmation in the show ring and their strength in pulling contests. At the first fair, Percherons, Clydesdales, English Shire, and grade draft stallions and mares showed in the draft horse classes. One of the "most elaborate and important displays" of the 1902 fair was the "heavy draft and coach horses." In 1915, Adolphus Busch III of Grant Farm in St. Louis brought a team of sixteen horses to the fair. According to the *Sedalia Democrat*, they attracted "much attention and admiration, as they rushed through the paces on the tan bark of the livestock pavilion at the night horse show." Used into the early twentieth century for pulling heavy wagons, construction work, and hauling heavy loads, draft horses remained in common use through World War I. As the numbers of tractors on Missouri farms increased during the 1920s and 1930s, the numbers of draft horses declined. After World War II, draft horses, though not used widely for farming, remained popular among enthusiasts and were used by Amish and Mennonite farmers who continued to rely on horsepower. In 1955, the Missouri State Fair offered $670 in premiums in a pulling contest. Between

fifteen and twenty teams entered. Winning teams in 1954 had pulled between 268 and 250 percent of their body weight, between 4,000 and 5,500 pounds, and the *Sedalia Democrat* predicted an exciting contest. Following a renewal of interest in draft horses in the 1960s, entries in draft horse shows increased and other breeds were added to the shows, including Clydesdale-Shire, Percheron, Belgian, and Haflinger halter classes and draft hitches pulling carts.

In addition to competition based on strength, teams of well-trained draft horses have put on demonstrations highlighting their horses and their elaborate trappings. In 1925, Billy Wales drove a six-horse team of Clydesdales from Union Stock Yards and Transit Company of Chicago. The most well known draft horses to perform at the Missouri State Fair are the Clydesdales owned by the Anheuser-Busch Company. In 1936, an eight horse hitch driven by Andrew Haxton visited from St. Louis. Haxton commented that "no horses in the country have finer trappings, nor more personal attention." The Budweiser Clydesdales appeared at the fair in 1941, and again in 1952, when they were valued at $50,000. The Budweiser Clydesdales returned to the fair in 1964, and several times through the 1970s, 1980s, and 1990s. A much smaller exhibition was presented in 1964 by the Victor Comptomoter Corporations' six-pony hitch of "matched Shetland ponies" who performed "precision maneuvers while pulling a miniature fire engine."

Rodeo

Just as the draft horse pulls grew out of contest of skill between farmers who used such horses each day, rodeo grew out of the skills needed to use horses in working cattle. The ability to separate one steer from a herd or a calf from its mother, the ability to rope and tie a calf, and the ability to tie a steer grew into competitions and professional organizations. The state fair rodeo, sanctioned by the Missouri Rodeo Cowboys Association, is held in the Dodge Arena on the western part of the fairgrounds. Here, cowboys com-

Young cowboys
rest on a fence.
(M. S. F.)

pete in calf roping, bull-dogging, bronco riding, and other events. Cowboys try to ride bucking horses or bulls with flair and finesse. Rodeo clowns cavort with the bulls as they protect the riders from injury. Cowgirls show their skill as riders and their horses' speed as they race around barrels. Rodeo fully demonstrates the ability of horse and rider to work together. Once again, as Clement predicted, the horse has been improved "in its susceptibility to take on education."

Though the horse is no longer the essential part of Missouri agriculture as it was in 1901, it remains an essential part of the lives of many Missourians, who love and cherish their animals. The horses' ability to work and to provide recreation guarantees that it will always be a part of the fair.

A Superintendent, an Exhibitor, and a Mule

The world of racing, horse shows, pulling contests, and rodeos could not exist without the management skills of department superintendents, the participation of exhibitors, and the cooperation of the animals.

LaRue Sauers was, with the assistance of his wife Dorothy, Superintendent of the Quarter-Horse Show at the Missouri State Fair for over twenty years. When the state fair horse show had become less an attraction, Sauers is credited with reviving the show during the term of fair Director Wilburt Askew and Commissioner of Agriculture Dexter Davis. For a period of time, Sauers had responsibility for five of the fair's horse shows. Horsemen and women describe him as an "outstanding leader" who was "highly respected."

The Roberts family of Clinton has a four-generation tie to the horse shows. "Uncle George" Roberts, Howard Roberts, Myra Roberts Finks, and her children continue to exhibit American Saddlebreds at the fair. "We look forward to the fair; it's the biggest show of the year for our family. It's been that way since I was a little girl," says Ms. Finks. Howard Roberts has been showing horses since he was eight years old; he's 85 now. He says that showing horses at the fair has improved. Equipment is better and the horses are finer, but he says many horses today don't have the stamina of yesterday's horses: "They were tougher then." The family has owned several prizewinners, including Easter Cloud, Easter Vanity, and Easter Serenade. In 1994, the family was honored for their participation in the Society Horse Show.

When President Taft visited the fair in 1911, he spent the afternoon golfing at the country club. There, he met A. J. Heck, who had originally built the fair's racetrack. As the men talked about the fair, Heck mentioned that he owned a mule that was about to foal. Taft suggested the mule be named for him. The mule was born, it was a male, and it was named Bill Taft. It grew to be a fine animal, valued at $400.

At the Races, State Fair, Sedalia, Mo.

CHAPTER 9

CAR RACES, MOTORCYCLE RACES, THRILL SHOWS AND TRACTOR PULLS:

The Checkered Flag

Not even cold and rain can keep race fans away, as these happy spectators demonstrate. (M. S. A.)

THE AUTOMOBILE WAS QUITE A CURIOSITY IN MISSOURI IN 1901. Very expensive, cars were status symbols few could afford. Missourians were very interested in this new technology, however, and wanted to see what the automobile could do. The first fair featured "a motor bicycle race and an automobile race and exhibition." The second fair did not have automobile races, but the 1903 fair again featured an exhibition race. During the first decade of the twentieth century, cars remained quite expensive. Lawrence

Christensen and Gary Kremer point out that a 1905 Studebaker cost $1,350, a figure that represented three years' wages for a factory worker. In 1908, Henry Ford introduced the Model T, a car that supposedly "everyone could afford," at a price of $825. However, prices continued to decline and by 1915, Kirkendall reports 76,000 cars in Missouri. By 1917, Model T prices had again dropped, this time to a more affordable $350, and by 1920, the number of cars in Missouri totaled 297,000.

A group of fair officials watch a race from a box seat. (M. S. A.)

Motorists were told that driving to the fair was easy. (M. S. F.)

The rapid growth of the automobile featured in both the state fair's advertising and its events. A 1921 advertisement extolled the ease with which one could "easily drive to the fair." In the 1930s, the Sedalia Chamber of Commerce offered free parking to fairgoers who would stay in downtown hotels and ride the streetcar to the fair each day. The fair added more automobile events to its schedule. In 1911, the area under the grandstand, formerly used for concessions, had been paved and was used as a demonstration area for car dealers. In 1915, Commissioner of Agriculture Jewell Mayes praised "the display of farm implements, motor cars, and expensive road building machinery" as being "the biggest in fifteen years of the fair." That same year, five Maxwell Automobiles sold at an auction on the grounds for between $560 and $591. The Automobile Fashion Show remained a popular event, and in 1917, the fashion parade was introduced as new model cars, driven by "Missouri's fairest women" dressed in the "most attractive garbs and creations of Dame Fashion" drove the cars in an "Auto Fashion parade."

A more exciting event, Auto Polo, was an international sport. Polo, known as a dangerous sport played on horseback by the rich and famous, had been adapted to the car. In 1921, a match was scheduled between American Team under leadership of Captain Eddie Buckern and Malletman Ray Reynolds and English team under leadership of Captain Elmer Robbins, and Malletman Jay Costello. The *Sedalia Democrat* suggested in 1922 that poor drivers might find the auto polo matches educational: "Automobile drivers . . . that don't know just how to handle their machines when they turn turtle or two or three wheels happen to break at the same instance can profit heavily by watching auto polo at the State Fair." When a car turns over, "they right the machine and then continue to battle for a goal."

The increasing number of cars created problems for Missouri. Roads were in poor condition, accidents occurred, and cars racing down roads and highways placed drivers and passersby in danger. Law enforcement agencies warned that "no road racing" would be allowed in Missouri, but automobile

Some of the biggest names in racing participated at the Missouri State Fair; drivers were identified by names on advertising posters. (M. S. A.)

Below: Auto Polo was a rough game, definitely not for the fearful. (M. S. F.)

owners, anxious to see what their vehicles could do, raced anyhow. Some of the problem of "road racing" was solved when races moved to the existing tracks at county and state fairs, and when racing became an organized sport.

The main racing circuit in the early twentieth century was the American Automobile Association circuit that raced on paved tracks. The AAA Circuit Races requested entry fees which many drivers thought unreasonable. A second racing circuit, the International Motor Contest Association, operated races on dirt tracks. The IMCA racers were, according to racing historian Ernie Roberson, considered "outlaws." The Missouri State Fair contracted with Alexander Sloan of the IMAC to provide races. By 1913, the fair advertised "Automobile and Motorcycle Races," along with "Aeroplane Flights and Races."

Although races had been held almost every year since the fair's beginning, *Kansas City Star* reporter Jim Conaway identifies 1915 as the beginning of State Fair Speedway. Newspaper reporters expected that "all dirt track speed records will be crushed" when Sloan brought "a string of racing cars" to the state fair. The "World's Speediest Drivers" were in Sedalia, including Mrs. Joan Newton Cunco, the "first woman to enter motor racing game." She was to be given $5,000 if she broke any records established by the male racers in the group.

The races were held on dirt tracks prepared for and used by horses. Drivers called "dirt track racers" raced cars called "dirt track cars" or "race cars." These cars had narrow tires which created ruts in the tracks, so racing on dirt was "like racing on a plowed field," according to local driver Bill Utz. The tracks were, says Roberson, "exceptionally dangerous." World champion racers appeared at the fair that year, including Johnny Raimey, short distance champion of the world, Eddie Hearne, and Louis Disbrow, the "prince of racers" who drove a Jay-Eye-See car. Disbrow announced that he would attempt to "make three miles in a minute from a flying start" on the fair's oval track.

The 1915 races also included a one-hour contest for speed and endurance. More than fifteen cars, driven by "the most daring and at the same time the most reckless" drivers would compete in the "most severe and trying competition that has ever been seen on any mile track." The local press promised fans would see "smashups and wrecks." The purses seem exceptionally large, considering the value of a dollar at that time. In the five-mile race, with a $500 purse, Louis Disbrow of New York raced against Eddie Hearne of Chicago. Hearne won with a time of 4:46.5. In an exhibition race, Louis Disbrow, driving a Case automobile, covered a mile in 54.5. Four cars participated in the Australian Pursuit Handicap, a five-mile race. Women's races were a regular part of the IMCA program. Bunny Thornton beat Elfredia Mais in the two-mile race with a time of 2:24.4. In 1929, women participated in a Society Automobile Driving Show. Mrs. F. R. Quint drove her Hupmobile Straight eight to a victory in division one, featuring cars valued at more that $1,200, and Miss Lady Judd Golliday's DeSoto won first place in division two, cars costing under $1,200.

In 1917, car races "will furnish the thrills for those who seek speed and endurance in motors tuned to develop both and who wish to see the most daring of drivers," stated the *Premium List*. The track provided "unequaled facilities for dirt track driving." Conaway suggests the races were less than had been anticipated; he cites a *Sedalia Democrat* reporter's comment on the feature race as "so bad that no one deserved to be recognized as the feature race winner." In 1918, World's Dirt Track Champions Earl Cooper, Louis Disbrow, and Gaston Chevrolet raced on August 12. Seven big events, constituting a three hour program, featured "seven of America's crack speed kings" The press advertised both drivers and their cars: Earl Cooper drove a Stutz; Louis Disbrow drove a Fiat; D'Alleen drove a Marmon; Gaston Chevrolet drove a Chevrolet Special; Jerry Wonderlich drove a Marquette Buick; Cliff Woodbury drove a Dusenberg; and J. Durant drove a Sunbeam. Louis Disbrow also raced against

Below: Spectators and pit crewmembers stand along the track in this 1930 photo. (M. S. A.)

Bottom: Grandstand and bleachers are full in anticipation of the race. (M. S. A.)

"Big cars" race in 1945.
(M. S. A.)

Right: Race advertisements often identified both drivers and their vehicles; this race was part of the state centennial celebration in 1921.
(M. S. F.)

Gertson, the "Human Night Hawk" in a demonstration contest between a car and an airplane.

The exaggerated advertising was part of the racing mileau. The 1921 races advertised the fair as "The Shrine of Speed" where thirty drivers were "hurtling past the judges and dashing just as madly round the curves, gambling with Death at every turn of the wheel." That year the cars included a Premocar, an Essex, a Chalmers, a Wyllis, a Peugot, a Severin, a Cyclone, a Rajo, a Hudson, and a C.R.G. The races were five miles and ten miles, from a rolling start. In 1922 Southwestern Bell had installed a "mammoth loud speaker." From this system, announcers broadcast what historian Economaki calls a "hyped-up spiel to get people on the fairgrounds to the track." The announcers, Conaway indicates, were "masters at what they did." The 1922 races were the "greatest automotive program ever arranged." The Fair Association had spent "thousand of dollars getting together this gigantic outdoor affair." Racing at speeds of over one mile per minute, the winner of the fifteen mile race winner boasted a time of 14:44.

Racing was a "young man's sport," commented the *Sedalia Democrat*, noting that all IMCA racers were under 29: "Mental and physical demands are so heavy and exacting upon the members of the dirt track division of the gasoline sport that usually a driver jumps from obscurity in a short time, basks in the spotlight for a few years and then retires, as they are unable to continue at this dangerous pastime, which always calls for a clear and steady hand."

The state fair did not hold auto races in 1923, 1927, 1930, and 1931. In 1924, Johnny Waters broke a track record, covering the mile

MISSOURI CENTENNIAL EXPOSITION and STATE FAIR

Auto Race Drivers Tempt Fate For Gold and Glory

You never saw such an exhibition of speed as is coming to Sedalia this year. Thirty of the fastest speed demons in the world will match their skill and their valuable speed monsters for the $5,000 prizes offered by the management.

It will be a great assemblage of racers when these panting, roaring speed monsters, piloted by reckless pilots, draw up before the many thousands of eager spectators in the Amphitheater on the Fair Grounds at Sedalia.

Here is a list of the famous all-star drivers who have entered their cars in the speed events:

"Bob" Robinson, a former Missourian, who holds a lot of records with his valuable Chalmers.

Larry D. Stone, who is known as "the World's Greatest Speed King," will pilot his famous Duesenberg around the mile track at Sedalia. He is driving the Duesenberg with which he made world's records at Des Moines in 1918.

Jimmy Costa, "Wild Jimmie," who lowered by three seconds the famous records of De Palma and Chevrolet in 1920, has entered his famous Fiat.

Lou Schiebel, "Central States Champion," will try for a new record with his wonderful Premocar.

In addition to these demon stars, there will be:

Joe Morino	Morino Special 30	Johnny Mercer	Mercier Special 13
Harry Delong	Severin Special 34	Johnny Waters	Premo Car 7
Swede Anderson	Chevrolet 15	Fich Seip	Hudson 16
Jan Boyd	Hudson 9	Antonio Bertillo	Ripon 18
Tuck Fordyce	Mercer 4	Luke Jobe	Stutz 8
Dick Dixon	Packard 12	Fred Lentz	Hudson 11
Earnie Frosnaugh	Peugeot 20	Henry Hughes	Hudson 25
Charlie Searles	Rajo 7	Chas. Kinney, Chalmers Blue Bird 3	
Joe Chatburn	Green Dayton 1	L. D. Stone	Benz 1
Eddie Buchanan	Green Dayton 2	Art Clark	Drexel 22
Jimmy Gray	Fiat 28	Ted LeDaro	Duesenberg 10
Raymond Roundtree	Essex 4	Carl Hauser	Marmon 18
Leon Delvaux	Minerva 10	Louis Betz	Mercer 14

Reduced Railroad Rates

MISSOURI CENTENNIAL EXPOSITION and STATE FAIR

Sedalia - August 8-20

EARL COOPER
LOUIS DISBROW
GASTON CHEVROLET

World's Dirt Track Champions with several
other well-known drivers will race at the

Missouri State Fair
Monday, August 12

track in 49.6 seconds. His record held only five years, for in 1929, Sam Hoffman, driving a Fronty-Ford, established a new record of 48.2. The Fronty-Ford was a popular car that year; Hoffman, second-place winner Tootz Campo, and Cannon Bill Baker Jr. also drove Fronty-Fords. In the Grand Sweepstakes, a fifty-mile race, four cars entered; two finished. Hoffman won, and a team of Campo and Shinny Jones finished second. During the 1930s, Gus Schrader, the "outlaw king of dirt tracks" dominated "big car" racing. Schrader raced at the state fair in 1930 and between 1933 and 1941. Fair Secretary Charles Green negotiated in 1933 with the IMCA to "stage a series of speed classics over the state fair oval," making Sedalia one of thirty-five cities to host IMCA races. Fifty of the world's leading speed stars, including Verne Ellis, the 1932 Southwest track champ, Rex Edmunds, and Emory Collins, the Canadian title holder, participated.

Dusenbergs, Frontenacs, Rileys, McDowells, Millers, Cragars, Dreyers, and Hispanos raced in the 1935 events. Twenty-four drivers raced in the National Circuit Championship; Sir Malcolm Campbell, world's straightaway record holder, Lou Schneider, an Indianapolis champ, and Schrader competed. Schrader, sometimes called the "Flying Dutchman," drove a Miller special equipped with Montgomery Ward Riverside tires, for which he had an endorsement contract. In 1936, on a "track dry but a bit rough," Posey

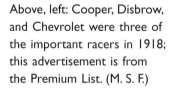

Reeves crashed into the inside rail on the south curve. He was not hurt, and his car only slightly damaged. Pete Alberni, the Italian racer who drove a car owned by St. Louis Cardinal Pepper Martin, provided spectators a glimpse of "sensational skidding which saw his car crosswise of the

Above, left: Cooper, Disbrow, and Chevrolet were three of the important racers in 1918; this advertisement is from the Premium List. (M. S. F.)

Above: Raymond Roundtree also raced during the 1920s. (M. S. A.)

At left: Larry Stone sometimes raced in a Dusenberg and sometimes in a Benz. (M. S. A.)

Jimmy Costa raced during the 1920s. (M. S. A.)

Bottom: Joie Chitwood raced and later opened a thrill show that performed at the fair. (M. S. F.)

track, and he drove almost the entire route of the 10-mile closing event with a flapping hood on his machine threatening trouble." "Gallopin' Gus" Schrader and Jimmie Wilburn, the world's banked track champion, drove "big displacement Offenhauser machines, Swan Peterson a stagger-valved Frontenac, and Tommy Kristin a Hispano-Suiza" in the 1941 state fair races. Just two months later, in October 1941, Schrader died in a horrific accident at the Louisiana State Fair in Shreveport, marking the end of a racing era.

Racing was prohibited by law during World War II, when all the United States' resources were diverted to the war effort. In 1946, racing at the fair resumed. In 1947, Jimmy Wilburn cracked Schrader's record, and ran a mile in 38.85 seconds. Two new features had been added in 1949—a one-hundred lap stock car race and an IMCA sprint-car race on the half-mile track. Though stock car races, involving cars people generally drove on the street, had been held at the fairgrounds in 1929, stock car racing is said by NASCAR historians, to have developed in the late 1940s, when Bill France Sr. proposed racing the sedans that people generally drove on the streets. In 1951, the fair used the one-mile track for the "stock car" race of one hundred miles in "late model

passenger autos." The fair also held the "big car" races on the mile track. "Big car," also called a "sprint car" by the 1950s, was a term used to distinguish the midget cars from the sprint cars. "Fans wanted to see Frankie Luptow's Offy, called the Black Panther, tested on the big track," wrote the *Sedalia Democrat*. Many of the cars then used airplane engines and "lack quick acceleration needed for the half-mile track but are noted for tremendous speed once they are rolling." The Offenhauser, with an engine developed by Harry Miller, was a popular racecar of the time. At the 1952 state fair races, nine of the contestants drove Offenhausers, including Bates City farmer and racer Jimmy Campbell, whose "geranium pink Offy" was instantly recognizable.

In 1955 the Missouri Modified Stock Car race, limited to the first fifty contestants to register, ran on the revamped half-mile track. The fair used the term "Jalopy" to describe the stock cars that raced on a banked, one-half mile track. Torch Aleshue, the "bad boy of the Jalopy Racing Cars throughout Missouri" won the twenty-five-lap race. That year, races were spread out throughout fair instead of being concentrated on one day. Bobby Grim, the leader in point standings, won the twenty-lap race. Speedway cars raced in a one hundred-mile classic on a one-mile track. Twelve thousand viewers watched. The press still used the terms "big car," and "jalopy" to describe types of cars, but more precise terms and more specific standards for cars were evolving. During the late 1950s and 1960s, IMCA-sanctioned sprint car races, midget car races, and modified stock car races were held. Late model stocks took to the one-half mile track for the first time in 1962 to the approval of an "enthusiastic crowd." The *Sedalia Democrat* attributed their popularity to "the close competition on the smaller oval." Races were so popular that in 1964, car races were reported to be the "only moneymakers in the grandstand." Racing greats such as Al and Bobby Unser, Jerry Blundy, Pete Folse, Arnie Knepper, and Bobby Grim raced at the state fair. In 1970, Al Unser placed first in the one hundred-mile race on the one-mile track with a speed of 98.039 miles per hour. Racing reached a high point in 1973 when ten different events were scheduled. By 1976, the fair offered six racing events with a total purse of $35,000. IMCA sprints, new model stock cars, and Missouri modified stock cars ran that year; a sprint car futurity highlighted the racing program. Sedalia driver Bill Utz remembers the Missouri State Super Modified races would "fill the grandstands, and the bleachers would be totally full, and they'd have standing room only. Even on the mile races, they'd fill the thing."

Left: Frank Luptow was the 1950 Dirt Track Champion; he often ran Offenhauser engines. (M. S. A.)

Below: Programs identify racers and their cars. (V. B.)

Bottom: Stock cars race in 1957. (M. S. A.)

"Big cars" go into the turn in a 1952 race. (M. S. A.)

Below, right: Gary Scott spins in his sprint car. (M. S. A.)

The 1977 races marked the arrival of a group including Sonny Smyser, Steve Kinser, and Doug Wolfgang, who, like Gus Schrader, would come to be called "outlaws." The drivers of the World of Outlaws, organized in 1978, wowed fans with their speed and daring at the 1984 and 1985 fairs. Winged sprints, sprint cars with a "wing" mounted above the car for stability, raced in 1979 and continue to run at State Fair Speedway. In 1980, midgets, a smaller version of the sprint cars popular in the 1930s and experiencing a resurgence in popularity, came to Sedalia. In 1997, after a season-long absence, the late model cars returned under the sanction of the Midwest Late Model Racing Association. The sprint car races that year were, according to the *Sedalia Democrat*, interesting: "The full moon . . . should have been the first clue" wrote reporter John Reidy. "The fans got a lot of excitement that night," he continued, "with spectacular crashes and a B feature that never reached its conclusion." Super modifieds, early model stocks, and late model stocks, as well as sprint cars, each having strict specifications for engine size, fuel type, and body design, race at the fair for the Missouri State Championship.

Racing is dangerous. Bumps, crashes, and spinouts are simply part of the sport. Most of the severe accidents are described by the press as "freaks," as mishaps that move beyond what is expected. In 1963, such an accident claimed the life of Bill Morris Billingsly of Oklahoma City. His car, "went into a cartwheel as it entered the third curve, and bounced over another car driven by Walter J. Wyrembeck. Billingsly's car did fifteen cartwheels, two of them in the air." An even more horrifying accident occurred in 1965, when four people were killed and fourteen were hurt during the one hundred-mile stock car race. A crowd of ten thousand spectators in the grand-

stand witnessed the crash. Driver Bill Crane lost control of his car in a blowout as it entered the first turn on the south side of the grandstand. The car crashed through the chain link fence into the crowd. Only a "few seconds before the accident," reported the *Sedalia Democrat*, a Highway Patrol trooper issued warnings over a loud speaker for spectators along that section of the fence to leave. He said they were in a restricted area." The spectators, which included several parents and their children, had climbed under one fence in order to stand at the fence enclosing the track. Other patrolmen were moving toward the crowd to order them back when the car hurtled through the fence. Trooper W. D. Ryan, who had served on the Missouri Highway Patrol for nine years, described the scene as "the most horrible sight I've ever witnessed." The injured, treated at the scene by ambulance attendants, volunteer firefighters, and Highway Patrolmen, were taken to Bothwell Hospital, which called in extra nurses and physicians. Following the accident, the design of the fences was changed to prevent such an accident from happening again.

Sedalia's State Fair Speedway brings people to the community throughout the racing season for sprint car races and World of Outlaws special events. On Friday nights from May through September the roar of the engines and the commentary of the announcer over the loudspeaker let everyone know that tests of car and driver continue in Sedalia.

Thrill Shows

Those desiring more excitement than racing could provide had the opportunity to see "death-defying" stunts performed by "motor maniacs" in

Left: Bob May, Sammy Swindoll, and Doug Wolfgang pose after a race. (M. S. F.)

Above: The Missouri State Fair Queen congratulates a winner. (M. S. A.)

Below: Racing can be a dangerous sport; this pile-up occurred in a 1957 race. (M. S. F.)

automobile thrill shows. Cars jumped over parked cars, drove on two wheels, raced with airplanes, and crashed into one another. Thrill shows have been a part of many fairs, but the press releases for the Aut Swenson Daredevils Thrillcade in 1955 seem best to exemplify them all. A "Ride of Death" performed at only a few places on the tour involved a "two vehicle leap" as "one daredevil spins his machines through a triple loop the loop while the other plummets beneath him at a surprising rate of speed." Carlotta DeMille, the "only feminine battering ram" was "currently studying further death neighboring stunts" because a thrill seeking public has called "for more from Carlotta." Steeplechase races, in which men standing on top of the cars raced against one another over elevated ramps, were "much more spectacular and dangerous that the toga wearing devils [the Romans who invented chariot racing] ever imagined."

A sprint car winner poses with his car as his crew checks the tires. (M. S. F.)

Right: A skull and crossbones exemplifies the "death-defying jumps" that were part of a thrill show. (M. S. A.)

Below: A driver propels his car under the car jumping from ramp to ramp in this Tournament of Thrills. (M. S. A.)

Motorcycle Races

Historians continue to debate about the whens and whos of the invention of the motorcycle. Like the car, the motorcycle increased in popularity as its versatility became apparent. Motorcycle races were held at early fairs, ultimately replacing bicycle races. In the years after World War II, motorcycles became more popular. In 1955, the fair hosted Motorcycle Polo. If auto polo was, as the advertisements suggested, "not a sport for sissies," motorcycle polo was truly a sport for the daring or perhaps the reckless. The "fast, sharp-turning motorcycles" and the "swinging mallets" meant danger for the participants and excitement to the fans. Although motorcycle races had been held at the fair in the 1920s and 1930s, it was not until the 1950s that the American Motorcycle Association developed a circuit of sanctioned races held on county and state fair racetracks. By the 1960s, motorcycle races were a major form of entertainment. The races at the state fair were particularly well known; one Sedalia woman attending college in Kansas in the late 1960s remembers "meeting people from California and New York. They all knew about Sedalia because of the motorcycle races at the state fair." Racing became less widespread as the cost of racing increased and the size of purses did not increase at a

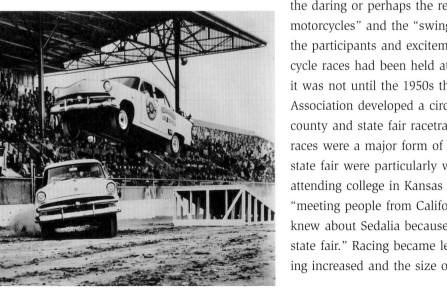

In a 1952 race, motorcycles race into the turn. (M. S. A.)

Medical personnel wait at the crash scene. (M. S. F.)

Bottom: Rick Yeager raced motorcycles; now his dealership, Yeager Motorcycle Sales, promotes races. (M. S. A.)

A fair queen poses on a tractor with tractor pull winners. (M. S. A.)

Some tractors look more like dragsters than tractors. (M. S. A.)

comparable rate; however, fans continued to enjoy the thrill of competition. Since 1993, Yeager's Cycle Sales in Sedalia has promoted the motorcycle races at the Missouri State Fair. The quality of the track is a drawing card for racers, and the quality of fair personnel a draw for promoters.

Tractor Pulls

Tractor pulls grew out of the draft horse pulls and plowing contests that marked nineteenth century fairs. As tractors became more widely used, farmers challenged one another to test the capability of their machines and their skills as drivers. The state fair and the *Missouri Ruralist* in 1951 formalized farmer and tractor competition with a Tractor Rodeo. This contest pitted winners of district contests against one another in events such as backing and docking a tractor and wagon, belting a machine to the PTO mechanism, and taking the tractor through a driving pen.

The first recorded tractor pulls took place in 1929, and they remained popular events at county fairs. In 1969 a National Tractor Pullers Association was established, setting uniform rules for pulls, and in 1970, the Missouri State Fair held its first tractor pull. What had begun as contests between standard tractors came to include modifieds, powered with "non-tractor type engines." Super stock tractors developed as pullers added turbochargers to stock tractors. As the tractors became more and more elaborate, interest in the sport "rocketed." The National Tractor Pull Association added a four-wheel-

Antique tractors are exhibited by FFA students.
(M. S. F.)

drive tractors division in 1976. As the tractors became more powerful, the
weight pulled by the tractor, called a sled, also changed. Originally a dead-
weight, then a sled on which people stood to add weight, and it now is a
"marvel of engineering technology, with sophisticated gearing systems that
move up to 65,000 pounds of weight." The tractor pulls at the 1991 fair featured contests between antique fuel powered tractors manufactured from the
turn of the century to 1939, and classic tractors made between 1939 and 1958;
242 tractors participated. The tractor pulls feature old super stock and modified
tractors that resemble dragsters as well as old tractors that resemble the ones
that grandpa drove.

The desire for speed and power expands with our capability to move faster
and move more weight. The Missouri State Fair, with its long history of racing
and pulling contests, has encouraged that desire among Missourians.

CHAPTER 10

"STEP RIGHT UP, FOLKS!":
Concessions, Carnivals, and Entertainment

I N DESCRIBING THE RECREATIONAL ASPECT OF FAIRS, HISTORIAN Wayne Campbell Neely recognizes that in "the racing, the vaudeville performances, the athletic events, the music and parading—there is something of the freedom of the out-of-doors, the wonder of skillful feats and courageous accomplishments, the thrill of close-fought contests, the expansive communion with a crowd." The Missouri State Fair certainly provides entertainment as part of the fair experience. From a well-balanced meal to a junk food binge or from Grand Opera to the Grand Old Opry, the fair has provided a range of concessions and entertainment. The changes in the type and method of food service, like the changes in the type of entertainment and the events of the carnival, reflect the changes in attitudes and lifestyle over the years.

Through the years, visitors have collected pennants, caps, matchbooks, and ticket stubs to remind themselves of their trip to the fair. (V. B.)

Concessions

Attending the "People's School," as the early state fair were sometimes called, was hungry work, and those attending needed sustenance. Before the first fair, the *Sedalia Democrat* called for food more substantial than the pies and cakes many would be serving. Substantial food did appear, as entrepreneurs established concession stands serving meals. Not so substantial food also appeared, as popcorn, peanuts, and ice cream tempted visitors. Later, stands serving tasty but less nutritious food materialized to satisfy the desires of those wanting what could be called "fair food"—cotton candy, candy apples, corn dogs, and pineapple whip.

During the first few years of the fair, downtown Sedalia was heavily involved in the fair. Out-of-town visitors, for the most part, reached Sedalia by rail. Sedalia's hotels were located downtown, and visitors could take a streetcar or the Katy train to the fairgrounds. After spending the day at the fair, they could return to downtown Sedalia and eat in one of Sedalia's thirteen downtown restaurants, or they could eat a home-cooked meal prepared by the ladies of Sacred Heart Church or Calvary Episcopal Church. Those desiring alcoholic beverages, then not sold on the fairgrounds, could patronize Sedalia's twenty-two downtown saloons.

Many visitors planned to picnic on the grounds, bringing a basket filled with goodies from home. The fairgrounds' park-like setting with shade trees and flowers made dinner on the ground an attractive and more economical option. The designation of several acres of the grounds as a camp site called White City, and the fair's renting of tents, cots, and other camping needs, enabled many to come to Sedalia by train or by car, rent what they needed to set up camp, and to cook their own meals.

Most Sedalia visitors took meals at home. Eating in restaurants was considered by many to be a frivolous and costly activity to be indulged in only by travelers or the wealthy. Repeated comments from the *Sedalia Democrat* about the number of visitors "crowding the turnstiles after the dinner hour" seems to suggest that afternoon was the most popular time to visit the fair.

Still, fair directors sought concessionaires willing to serve food at the fair. Board minutes reveal that in January 1903, Dan Crow, Frank White, and Ira Bronson asked permission to "erect an eating house" in a design to be approved by architect Thomas Bast. The cost of a license to "conduct an eating house" was $25. Bast, grounds superintendent and superintendent of privileges, sold licenses to those wishing to serve food or conduct "general stands" selling lemonade, ice cream, candies, and popcorn. Those selling snacks and beverages, for the most part, did quite well. Bast, with the approval of the Board of Directors, regulated the concessionaires. For example, in 1906, the

Crowds congregate on the Midway, seeking the excitement of the hour. (M. S. A.)

A trip to the fair has to include a corn dog with mustard. (M. S. F.)

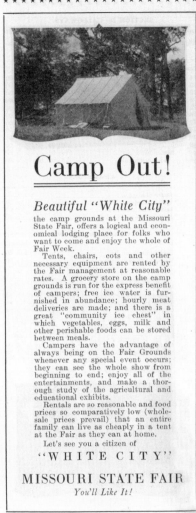

Camp Out!

Beautiful "White City"

the camp grounds at the Missouri State Fair, offers a logical and economical lodging place for folks who want to come and enjoy the whole of Fair Week.

Tents, chairs, cots and other necessary equipment are rented by the Fair management at reasonable rates. A grocery store on the camp grounds is run for the express benefit of campers; free ice water is furnished in abundance; hourly meat deliveries are made; and there is a great "community ice chest" in which vegetables, eggs, milk and other perishable foods can be stored between meals.

Campers have the advantage of always being on the Fair Grounds whenever any special event occurs; they can see the whole show from beginning to end; enjoy all of the entertainments, and make a thorough study of the agricultural and educational exhibits.

Rentals are so reasonable and food prices so comparatively low (wholesale prices prevail) that an entire family can live as cheaply in a tent at the Fair as they can at home.

Let's see you a citizen of

"WHITE CITY"

MISSOURI STATE FAIR

You'll Like It!

Page 92

The fairgrounds campground was called White City because of the color of the tents. (M. S. F.)

Right: Peter Pehl managed one of Sedalia's finest restaurants. (S. P. L.)

Board minutes note that a concessionaire "charging exorbitant prices was ordered to be excluded from the grounds." Further concern about prices is reflected in a 1910 order that concessionaires "post prices in large type" on the front of their buildings. Fair directors had other concerns, one of which was food safety. Passage of the nation's Pure Food and Drug Act in 1906 caused people to be aware of food safety, and in 1911, concession stands were inspected to be sure properly inspected meat and unadulterated foods were being served. A stand's appearance was also a concern; in 1918, the Board reported complaints about "unattractive booths and tents put up by concessionaires."

Board minutes indicate that in 1918, cider, fruit, and watermelon concessions operated on the grounds. The 1930 *Program Book* identifies juice, watermelon, and fruit concessions, but also mentions ice cream, ice cream sandwiches, caramel popcorn and peanuts, soft drinks, and root beer being served. In addition, Anna Kahler operated a dining hall and the Smithton Methodist Episcopal Church and the Epworth Methodist Episcopal Church had food stands. While the early years of the Great Depression hurt fair attendance and concessionaires' profits, by 1935, the *Sedalia Democrat* could report that "business is on the mend." The next year, "probably twice the number of concession stands" appeared on the grounds, further indication of the recovery of the economy. The extremely hot weather during the 1936 fair pleased concessionaires, who saw an increased demand for cold drinks and ice cream.

The 1950s ushered in an era of prosperity to the nation, and the widespread use of the automobile made Americans more mobile. As a result, people were more likely to eat away from home. The increasing number of food vendors on the fairgrounds confirms this change in attitude, as does the variety of foods offered. Stands serving meals "kept us hopping" said a local woman, Ida Shobe, who worked as a waitress in one concession. "We served fried chicken, pork chops, potato salad, peach cobbler, and were always busy." Church ladies still served plenty of homemade pies, though Missouri Food and Drug Laws prohibited "custard pies or other baked or cooked foods containing whipped cream or custard cream."

STYLES OF TENTS FOR RENT

No. 3 Baker Family Tent, with 6 foot walls and 2 awnings; the back side of the tent will be like the front side shown in the illustration.................... $10.00
12x19 feet, with 3 rooms.................... 12.50
14x24 feet, with 5 rooms....................

No. 2   Chautauqua Family Tent, with 6 foot walls, with awning.......................... $6.00
10x14 feet, with 2 rooms.................... 7.00
12x14 feet, with 2 rooms.................... 9.00
12x16 feet, with 3 rooms....................

No. 1   Plain Wall Tents................... $4.50
10x12 feet, each.......................... 5.00
12x14 feet, each.......................... 5.50
14x16 feet, each..........................

EQUIPMENT
Folding chairs, each...................... $ .15
Folding Cots, each........................ .75
Double blankets, each..................... 1.00
Double sheets, each....................... .40
Feather pillows with slip, each........... .35

Page 191

Left: Fairgoers could rent tents and other camping equipment. (M. S. F.)

Above: A family enjoys a picnic on the grounds. (S. D.)

Below: Visitors enjoy sandwiches and soft drinks. (M. S. A.)

This 1941 concessionaire seems happy to be at the fair. (V. B.)

Below: Concession stands line the street to the Midway. (M.S.A.)

Food and Drug Laws also governed the sale of fair staples—hot dogs, snow cones, and ice cream. Meats had to be purchased from an "approved source" and free from "extenders" and "chemicals." Another regulation allowed the sale of "ice confections made by pouring syrup over crushed ice," if the ice was "from an approved source and the syrup had been sterilized." Even more stringent regulations covered pineapple whip. Not only did the machines that made the product have to be approved in advance, the pineapple whip had to be labeled as a "frozen dessert" if its butterfat content was below eight percent, the appropriate amount for "ice cream." Advertising remains a concern: Lemonade must be made with lemons.

Almost anything a fairgoer might want to eat can be purchased on the grounds. Steaks, pork chops, and smoked turkey drumsticks and fried chicken dinners are options for those who wish to sit down and dine. Ham and biscuit sandwiches and summer sausage sticks offer the best of Missouri's meat curing companies. Corn dogs, hot dogs, and hamburgers, spiral cut French-fried potatoes, and onion rings please the palates of fast food junkies. Kettle corn, caramel corn, candy apples, caramel apples, cotton candy, and pineapple whip, those staples of fair food, are available on every corner.

Carnival

The concept of the Midway as a separate area for amusements and shows developed at the Chicago World's Fair in 1893 and quickly spread to state and county fairs. The first Missouri State Fair did not have a Midway; instead, the Elks Club hosted a street fair in downtown Sedalia. The street fair soon disappeared, and its "Streets of Cairo" exhibit was eclipsed by the activities on the fairgrounds. In 1909, for example, the fair hosted an Igorrote village featuring people from the Philippine Islands, "living here just as they lived at home."

Such exhibits were said to be educational, to allow Missourians a glimpse of life in a distant land, but the sign advertising the village on the fairgrounds suggests more than a bit of sensationalism: "Twenty-five Barbarians . . . dog eaters . . . head hunters . . . fully two thousand years backward on the scale of civilization." Other types of shows, especially "indecent and immoral exhibitions," were not permitted. Questions were frequently raised about what was considered "indecent," as one Sedalian interviewed in 1977 remembered that her parents "wouldn't let her meander near the midway where, along with the Filipinos, lurked some less-enlightening entertainers such as snake charmers and shimmying belly dancers."

SOME OF THE JOHNNY JONES SHOWS

Page 234

Neely comments that "one expects to find at the fair countless forms of bizarre entertainment, extravagant and impossible exhibitions." The state fair's Board of Directors remained perplexed about what constituted appropriated entertainment. In 1903, the board tried to secure Ella Ewing, a "giantess" as an "outdoor attraction," but in 1906, rejected the application for booth space of a "freak woman and midget monstrosity." In early 1913, the Board issued a statement: we "do not believe that a big carnival company should have a place on the Missouri State Fairgrounds." By 1916, the Board agreed that the fair needed a carnival. Neely believes that "In a holiday mood, often away from home and therefore freed of many of the primary-group restraints, many easily succumb to the lure of 'trying their luck,' testing their strength, treating themselves to the 'educational opportunities' offered by two-headed calves, pig-faced boys and Egyptian mummies, or indulging in various intemperances of eating, drinking, and spending." The 1921 Johnny Jones Shows, one of many traveling carnivals that crisscrossed the United States during the 1920s, set up the Midway. It allowed visitors the chance to see "Trained Wild Animals" and "Egyptia," which featured a "garden of girls" and Egyptian dancers. During the 1930s, George Loos Shows provided shows and games. The Cetlin and Wilson World on Parade, which dominated the Midway

Left: Johnny Jones Shows offered a variety of entertainment options. (M. S. F.)

Pineapple whip and ice cream are staples of the Missouri State Fair. (R. C.)

Below: This postcard highlights the 1909 "educational exhibit." (M. S. F.)

The Honey Girls showed off their fur coats in their dances. (M. S. A.)

Below: The Red, White, and Blue chorus line added a patriotic flair. (M. S. A.)

from 1951 through 1957, boasted "twenty-two new and highly entertaining shows." One show, the Hi Frenchie Show with Sally Rand and other "beautiful girls," raised questions about the nature of its offerings until Governor Forrest Smith visited the show and announced that it was "neither immoral, lewd, or obscene. I thought the production well presented." Raynell Golden managed both minstrel shows and "hootchie-cootchie shows" for Cetlin and Wilson. Raynell's shows changed names frequently, giving fairgoers the impression of new and varied entertainment. She retitled the Hi Frenchie Revue as Manhattan on Parade and then as Hollywood on Parade. The carnival barker's hype was often more suggestive than the shows, as the 1955 "Raynell's Pink Garter Revue" with "exotic dances" and "Mitzi the Milk Bath Girl" indicated. Visitors may have been more shocked by Liska Siaha, the girl with green hair, and her trained macaw.

The barker's cry lured people to the freak show, which in 1951 provided glimpses of the "alligator skin boy, the human monkey girl, the fire eater, the human electric dynamo, and a sword swallower." Over the years of Cetlin and

Wilson's tenure, their freak show displayed Percilla the Monkey Girl, her husband Emmett the Alligator Boy, Lobster Boy, Curley the Painproof Man, Henry the Pinhead, Mildred the Iron Foot Marvel, Zoma the Jungle Girl, and a fat show in which all performers were obese. Gradually, however, the number and grotesquery of the freak shows diminished. People became more knowledgeable of medical problems and more sympathetic to the handicapped. Freak shows during the 1970s marketed themselves as "educational," claiming to reveal the tragic conse-

The Inferno lures visitors into a chamber of horrors. (M. S. A.)

Below: A young man won a prize and perhaps a lady's heart. (M. S. A.)

quences of drug use. As the number of shows decreased and as the fair began offering more entertainment at other places on the grounds, the number of games and rides on the Midway increased. Midway games have always involved some test of skill or of luck. Missouri's gambling laws forbade games of chance, and during the 1930s, a Skill-o game was closed because it was considered too much like the proscribed Bingo. Testing one's skill could be chancy for the player or for the game operator. At a fair in the late 1940s, a Knob Noster man was asked, finally, to stop playing a game involving pitching pennies onto plates, as he had won too many times.

Roller coaster designer John Wardley affirms that "there is a place in society for providing fun and thrills that are exhilarating and where there is a perceived sense of danger." Carnival rides offering speed, movement, and heights

confirm Wardley's understanding of our desire to be frightened—safely. The basic carnival rides developed gradually over the eighteenth and nineteenth centuries, becoming practical and widely used after the opening of amusement parks such as Coney Island and the beginning of the "Midway" at the 1893 Chicago World's Fair. The Ferris wheel, for instance, was invented by George W. Ferris for the 1893 World's Fair; the Eli Bridge Company developed a practical, portable

The penny arcade promises to tell these teens' future, their personalities, and their love lives. (M. S. A.)

A rider could see the entire fairgrounds from atop the double Ferris wheel. (V. B.)

wheel, enabling carnival companies to take Ferris wheels on the road. The carousel with hand-carved horses and other animals reached its peak in the early twentieth century when many skilled artisans emigrated from Europe and put their skills to work carving the merry-go-round's menagerie. The roller coaster, originally a ride down a rolling incline, was redesigned by LeMarcus Thompson and further refined by John Miller. Most carnival rides stem in some way from these rides.

At early fairs, adult swings and carousels offered slow and gentle rides, maintaining the decorum thought to be appropriate for respectable people. By 1942, attitudes toward carnival rides had loosened considerably, and the World of Today Shows brought "fifty-one big trucks heaped high with midway paraphernalia" to the grounds. Their rides included the "old favorite" Ferris wheel, a carousel, the Octopus, the Rotoplane, the double loop plane, a Tilt-a-Wheel, a Spit-fire, and children's rides such as scooters, baby autos and planes, and a hay ride. Cetlin and Wilson offered three ferris wheels, a merry-go-round, a Roll-o-plane, an Octopus, a Spit-fire, a Tilt-a-whirl, the Looper, a Caterpillar, dodgem cars, and a moon rocket. Their Kiddleland featured seven rides, including a train and live ponies.

Entertainment

The entertainment at early fairs generally involved music or

Left: This restored carousel features a complete menagerie, including a rooster. (T. P.)

Below: Children try to catch the brass ring on the merry-go-round. (M. S. A.)

Left: A rocket ride captured the lure of outer space for the pre-Sputnik generation. (M. S. A.)

Below: Little ones pretend to chase one another round the motorcycle track. (M. S. A.)

CANTORE   LAVITTO

Opera Singers

APPEARING WITH THE FAMOUS

THAVIU BAND

¶ The musical world knows no more famous baritone than Signor Cantore. Few lyric tenors are more popular than Signor Irving Lavitto. Both artists are far-famed for their rich voices and highly temperamental renderings of whatever songs they sing. Fairgoers will delight in their numbers.

HEAR THEM EVERY AFTERNOON AND EVENING *at the*

Missouri State Fair

Music You Can't Resist!

Page 122

Fairgoers in the 1920s enjoyed grand opera in the grandstand. (M. S. F.)

Right: The Murray Family Orchestra publicity photo appears to have been taken in the family's living room. (M. S. A.)

Inset: Mabel Murray soloed in the family's performances. (M. S. A.)

aviation. The music was vastly different from what is available at the fair today, and included opera and orchestra performances on stage, as well as marching bands parading through the fair and vaudeville acts, often interspersed between horse races. A 1913 *Sedalia Democrat* advertisement mentions only Thiviu, a musician who returned to the fair with his musical troupe each year until the mid-1930s. In 1919, for instance, the fair spent $3,700 for Thiviu's troupe, the most spent that year on any entertainment feature. Thiviu presented extravagant concerts with operatic arias, choir numbers, orchestra renditions of classical and popular songs, as well as ballet and interpretive dance. Thiviu devoted Sunday to sacred music; the 1915 fair, for example, featured Thiviu and his band, along with Sedalia's Ladies Musical Club and the Apollo Club with soloists Miss Rogers, soprano; Mr. Benigo, tenor; Herr Mark Bing, baritone; and A. F. Thiviu, cornet. Sacred concerts became a standard feature of the fair. In 1927, Thiviu presented *The Prodigal Son*, a sacred drama that had won first prize in a composition contest in Paris. "Fair visitors," said the *Sedalia Democrat*, "are fortunate in having an opportunity of seeing it." Other musicians participated as well. In 1923, the Boy Scout Bands of Springfield and Bolivar and the Mexican Police Band presented music before a worship service on the grounds. Patriotic concerts had become popular during World War I and remained so during the 1920s and 1930s, when patriotic songs, inspirational songs, hymns, and popular music were combined in the programs. In 1919, Kronke's Band, a well-known Sedalia orchestra, the Second Regiment Band, a "home guard" military band, and an African American band performed at the campgrounds, in the grandstand, in the Coliseum, and at other points on the grounds. These acts, plus the singing Thomas Sisters, cost the fair a total of $1,500.

Vaudeville, a term used to designate stage shows combining

continued on page 216

Sally Rand — Missouri Girl Comes Home, Shows All

Sally Rand gained fame—or notoriety—at the Chicago World's Fair in 1933. Wanting to be hired to dance in "The Streets of Paris Show," she managed to gain the attention of the show's director by being arrested four times in one day for impersonating Lady Godiva by riding through the fair grounds on a white horse. The director hired her, paid her $5,000 dollars per week, and she danced. She had a long career as a dancer, performing at the World's Fair in San Francisco in 1939. Her famous "fan dance," in which she covered herself, usually clad in a flesh-toned body stocking, with fans of ostrich plumes, and her "bubble dance," in which she hid herself behind large balloons, made her a figure to remember. When Sally Rand came to the Missouri State Fair in the 1950s, she was coming home.

Sally Rand, named Helen Beck Gould, was born in Elkton, Missouri, a small town in Hickory County. The family moved to Jackson County while Helen was young, and she began to dance in Kansas City Theatres and studied voice, drama, and dance. After a brief career in silent movies, she focused on exotic dance. In 1951, she became part of the Cetlin-Wilson Shows, a traveling show that played fairs along the East Coast. From 1951 through 1957, Cetlin-Wilson played the Missouri State Fair, and Rand performed. She "adapted well to the carnival grind," says historian Bob Goldsack, "generating reams of publicity as she willingly attended various luncheons, dinners, press receptions while granting numerous TV and radio interviews." She endeared herself to Missourians and fair officials with her good-natured willingness to give out trophies, pose with livestock, and talk with fair visitors. She is one of the most enduring and endearing memories of the Missouri State Fair.

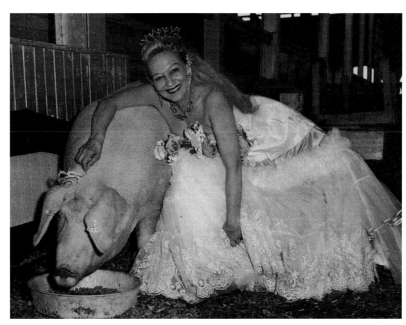

Sally Rand awards a tiara to a queenly hog. (M. S. A.)

Left: The Kawana Japs were a comedy juggling team, perhaps from Japan. (M. S. F.)

Right: Franklin and Astrid performed Mexican dances. (M. S. A.)

THOSE TOKIO TRICKSTERS
The Kawana Japs

UNIQUE entertainment is this, offered by two tiny artists from the Land of Cherry Blossoms. It consists of ground acrobatics, comedy barrel kicking, juggling, and a fascinating exhibition in dexterity—water spinning. Everybody likes to see "the Japs." This "Jap Act" is so decidedly different from the rest that it is sure to please.

MISSOURI STATE FAIR

Page 204

Left: Acrobats in scanty costumes were thought risque. (M. S. A.)

Above: A special ice arena had to be prepared for these skaters. (M. S. A.)

Right: Visitors enjoyed the smooth sounds of Hawaiian music. (M. S. A.)

Top left: The McCune Trio clowned, sang, and danced. (M. S. A.)

Top right: The Trio Angelos awed spectators with their aerial act. (M. S. A.)

Above left, and inset: Lester, Bill, and Gaines look quite different in street clothes than they do in their costumes. (M. S. A.)

Above: "Excess Baggage" told the story of two dogs on a trip. (M. S. A.)

Top: Jonny Rivers brought diving mules and a petting zoo to the fair. (M. S. A.)

Right: Griffith and Mike demonstrated the intelligence of the mule. (M. S. A.)

singers, dancers, acrobats, and comedy routines, also came to be a part of the fair. The vaudeville acts—the Caldanos, the LeMaze Brothers, the Lavini Trio, the Kawana Japs, the Steimer Trio, and Bob the Boxing Kangaroo—delighted the 1919 fair audiences with acrobatics, juggling, comedy, and singing. Acrobats and jugglers were especially popular, though female performers in tights were considered a bit risqué. The 1936 feature attraction, Ernie Young's Trip Around the World, offered various acts representing different cultures, and some of the performers actually came from the countries whose music and dance they performed. For example, Mexican dancers Franklin and Astrid were probably not from Mexico, though the Hawaiian Musicians may have been from Hawaii. The members of the troupe had been "gathered from the leading headliners of the vaudeville stage. A ballet and chorus has been recruited from the Hollywood studios and the many follies companies with the result that some of the prettiest and most talented dancers form this feature of the entertainment." The Trip Around the World was "embellished with a complete scenic equipment and so large an outfit it takes a 70 foot baggage car to transport the properties from fair to fair" and featured "electrical effects that turn night into day and rival the sun in brilliancy." A review of the show by the *Sedalia Democrat* mentions a two-hundred foot set, the Four Midnight Suns vocal and instrumental quartet, the DeKohl juggling troupe whose performers balanced on large globes as they moved up and down an incline, and a group of Swiss bell ringers, girls with bells on their ankles who played melodies by shaking their ankles, as well as a whirlwind roller skating act.

Vaudeville shows provided most of the entertainment at the fair through the 1940s and 1950s, though as movies and television came to replace live theatre, the term variety show was more likely to be used to describe the shows. The stage show and water carnival performed twice each night in front of the grandstand in 1951 included singing acts, interspersed with comedy. Ming and Ling sang hillbilly, Irish, and Scottish songs. Betty and Benny Fox, sky-dancers, danced on a very small platform atop a very tall pole. Trained dogs acted a skit called "Excess

Top: Walter Stanton trained a rooster to do tricks. (M. S. A.)

Middle: Bugger and Toots showed themselves as the world's largest steer and the world's smallest rider. (M. S. A.)

Bottom: Dolphins delight the young and old with their friendly personalities. (M. S. A.)

An Acrobat at the 1913 Fair

In August 16, 1913, *Collier's, The National Weekly* published Arthur Ruhl's account of his visit to a county fair, which he described as "one steadfast institution in a world of change." There he met a young acrobat with the traveling show: "I recognized the young equilibrist who, an hour before, in pink spangled tights, had been balancing on one hand on a pyramid of bottles. He had been assisted by a plump young woman in a white sweater, short skirt, and red stockings, with her hair, tied with a red ribbon, hanging down her back. It was her duty to bounce about the stage waving the traditional gestures of admiration and amazement, now and then pretending that she thought he was going to fall."

The young man, now dressed in his street clothes, discussed the life of a traveling performer: "They had their own tent and cooked their own meals and slept of fresh straw with a sheet spread over it. 'It's just like camping out. Why, we can keep out right up to winter. Next week we'll be over in Sedalia, Mo. . . .' It looked like a great life, I told him; the only trouble was how long could one keep it up? An acrobat had to be young. 'Keep it up? he demanded. 'Why, I know men sixty-five years old who do the most wonderful things you ever heard of. You've got to keep in shape, that's all—and that's easy enough, outdoors. Why, look at my wife. I took her out of a hospital—and look at her now!' There was certainly no doubt about her present good health."

The young couple enjoyed their life. They liked the travel, the camping, and the outdoor work of playing carnivals: "Everything seemed all right in a very good world to this young man, standing there with his husky arms folded across a deep chest, watching the crowd go by."

Ticket stubs remind visitors of concerts past. (V. B.)

Right: Ruth Law, a young flier, produced and participated in an airplane and automobile thrill show with Katherine Stinson. (M. S. A.)

Facing page: Gertson the Birdman, a barnstorming pilot, flew exhibition flights and raced with an automobile. (M. S. A.)

Baggage." The Variety Show format appeared again in 1955, when Barnes and Carruthers Stage Revues appeared in front of the grandstand. Their comedy routines, dancing, singing, music, novelty acts, animal acts, and clown BoBoBarnett were very popular with the fair's audiences.

Gradually, variety shows came to be replaced by concerts by individual artists, with different concerts each evening of the fair. Pop stars Donny and Marie Osmond, Tiffany, Bobby Goldsboro, and Christina Aguilera have highlighted fair entertainment bills, as have rhythm and blues groups the Jackson Five and Ike and Tina Turner, rockers Paul Revere and the Raiders, Grand Funk Railroad, R.E.O. Speedwagon, and Z.Z.Top, alternative rockers the Gin Blossoms and the GooGoo Dolls, and the legendary Bob Dylan. Most of the concerts at the fair, however, are country music. In 1935, the WLS Barn Dance appeared in the Grandstand. Stars such as Charley Pride, Barbara

Mandrell, Jim Stafford, Hank Williams Jr., Willie
Nelson, Waylon Jennings, Garth Brooks, Alan Jackson,
LeAnn Rimes, and Kenny Chesney have played the fair
before packed grandstands. Bob Hope and Bill Cosby
have entertained audiences with their comedy routines.

The other most popular type of entertainment over
the years at the fair has involved aviation. Balloon
ascensions, airship appearances, and barnstorming airplanes constituted the most spectacular entertainment
of the early twentieth century. Spectators at a 1915 balloon show received more excitement than they had
anticipated. Balloonist Johnny Mack invited Pearl Fay
Piercy to go up in balloon with him as a feature attraction. She was seated on a trapeze bar hanging below
the balloon when she slipped and dropped from the
bar. Her wristbands tangled on the trapeze and would
not allow her to open her parachute and float to the
ground. Instead, she hung on to the trapeze as the
balloon drifted farther and farther away, while the
desperate Mack tried to make the balloon descend.
According to the *Sedalia Democrat*, much of the crowd
followed the balloon off the fairgrounds into the coun-
tryside, where it finally came to rest three miles away.
No one was injured, and the crowd had a hair-raising
story to tell.

The Wright Brothers brought their plane in 1909
and again in 1910, when the Board of Directors hoped
to make $500 by charging an extra fee to see the plane
close up. Airplanes continued to be an attraction. Ruth
Law and Katherine Stinson appeared; fairgoers were
invited to "see them drop bombs on trenches, engage
in aerial warfare, and fight battles in the clouds." Billed
as "the world's Premier Aviatrices," the *Premium List*
bragged that "no flier on the French or Belgian front
can compare with these two brave young women for
daring feats." In 1919, the fair spent $2,100 for fliers
Louis Gertson, the Birdman, and Ralph Snavely.
Gertson often flew at night, with his plane trailing visible smoke that traced the loops and twirls he made as
he flew. As airplanes became less a novelty and more a
common means of transportation, aviation shows
became less often featured. They remained popular,
however, and the Golden Knights, an Army Paratrooper
Group, have performed intricate jumps.

A parachuter falls toward
the fairgrounds. (M. S. A.)

In a world perhaps jaded by the spectacular special effects of movies and
the constant presence of television, the gasps of awe from a crowd watching a
fireworks display and the waving lighters when Willie Nelson sings "Will the
Circle Be Unbroken" remind fairgoers that humans can still be brought
together by a sense of wonder at a state fair.

Cooking for Prizes and Profit

Maxine Griggs remembers the fair well: "It was always very important to our family." In 1938, her sister Lucille Bowers and husband Lloyd Bowers built the second permanent concession stand on the grounds. The stand, located between the mule barns and the Angus barn, originally had a canvas top supported by wooden posts cut by Lucille's father. Later the stand was a wooden building with a kerosene stove inside and a cook stove out back. Maxine, who began working with her sister when she was twelve, remembers carrying water from the Jersey barn and sleeping on a cot behind the stand.

In 1965, the family purchased six concession trailers. Maxine's daughter Judy Bell and husband Raymond continue to manage three corn dog concessions. In addition, Maxine's brother Lloyd Kindle has six concessions on the grounds. Her sister Nadine Klein cooks in the Youth Building cafeteria. Young people in her extended family, like Maxine, grow up working concessions at the fair.

Maxine, herself a restaurateur, prepared the Ham Breakfast for seventeen years, and prepared dinners for a number of celebrities, including Dottie West, Waylon Jennings, Willard Scott, Tammy Wynette, and President Ronald Reagan.

Maxine shines as a contestant, too. She has won ribbons for her displays of antique quilts and tablecloths, and in 2002 won first place with a ceramic rooster she glazed and fired with an unusual technique. But she is best known for her cooking. Over the years, she has won first prize for a variety of dishes. For five years, she placed first in the pork-cooking contest, and has won first place for her Spam recipes, for her berry pies, and first and second place for her chicken dishes. One chicken dish, which she says she invented because she had some leftovers that had to be used, merited not only a prize but also placement on the first page of a cookbook. In 2002, Maxine brought home thirty-nine ribbons—blues for a rabbit dish, for a turkey recipe, and for a rice dish; reds for chicken, and for Spam; and a white for her use of Pillsbury Pie Crust; as well as ribbons for a German Chocolate cake and for a cheesecake.

Maxine says she learned to cook by watching the good cooks she worked with both as a waitress and as part of the kitchen staff. She learned very well.

CHAPTER 11

"ONE HUNDRED FAIRS OF FUN":

Celebrate

The official logo celebrates the one hundredth fair. (M. S. F)

THE ONE HUNDREDTH MISSOURI STATE FAIR OPENED AUGUST 8, 2002, with balmy weather, the intense heat of the previous weeks softening into mild temperatures and cool breezes. Crowds gathered to see the best of Missouri. Governor Bob Holden formally opened the fair, telling visitors "to go out and have fun." The eleven day celebration of one hundred fairs concluded a year's worth of work by Department of Agriculture Director Lowell Mohler, Fair Director Mel Willard, department superintendents, state agencies, state fair staff, and various committees. Planning for the celebration began in 2001 with the selection of a centennial committee. Representatives from the

State Fair Commission, the Missouri Department of Agriculture, the Missouri Department of Natural Resources, the Missouri Conservation Commission, the State Historical Society of Missouri, *Missouri Life*, *Rural Missouri*, the University Extension Service, the Pettis County Historical Society, and the Sedalia Area Chamber of Commerce met several times; the committee brainstormed ideas that have governed the fair and activities that should be a part of the one hundredth fair. In a larger sense, however, the celebration of the one-hundredth fair was the culmination of all the fairs before. As the committee decided what souvenirs to sell, what events to host, what contests to hold, and what themes to empha-

Governor Bob Holden opens the one hundredth fair. (V. B.)

Below: Banners welcome visitors to the fair. (M. S. F.)

size, it returned again and again to previous fairs.

The fairgrounds itself looked different, shining and new but still historic. The vintage buildings, refurbished with new trim paint, now feature signs outlining their history. A new entrance off Highway 65 focuses attention on the new Centennial Entrance and the 1906 Coliseum. In the spring and summer, contractors demolished the dilapidated bleachers, removed the mile racetrack, and graded the banks of the former track. They repaired the Main Gate and relocated a maintenance shed. Over one thousand new parking places make access to the fair's attractions easier, and the reorientation of the grounds makes

it appear more spacious. Flowerbeds and boxes burst with color. The new Future Farmers of America Building and a new Highway Patrol headquarters show styles compatible with the existing buildings.

The fair opened on the morning of August 8 as exhibition buildings opened their doors for the crowds that thronged to see the latest in kitchen gadgets,

Above: 4-H members carry banners in honor of the 4-H centennial. (R. C.)

massage chairs, carpet cleaners, and holographic portraits. Barns opened to reveal the best of Missouri livestock, and judges began evaluating 4-H and FFA sheep, swine, Angus cattle, and poultry. At the Home Economics Building, judges sampled pasta salads, casseroles, and desserts. Art lovers wandered through the Fine Arts Building admiring the works that had been designated Best in Show. Governor Holden dedicated the Centennial Entrance to the grounds in the morning and Mrs. Holden the Museum in the basement of the Womans Building in the afternoon.

The state fair parade, a fixture at earlier fairs discontinued during the 1900s in part because of traffic congestion, once again marched through Sedalia. Sirens whooped. Drums beat a cadence. Flutes shrilled and trumpets sang. Old Glory snapped in the breeze. A vintage Highway patrol car, its single red light shining, led the way as the Centennial fair parade began. Marching bands, floats, clowns, forty John Deere tractors, an antique fire engine, and old trucks paraded south on State Fair Boulevard from Third Street into the Main Gate of the fairgrounds. 4-H Club members carried banners representing 4-H work in all 114 of Missouri's counties. Politicians, fair officials, and queens waved to the crowd from state fair trolleys, vintage cars, and a Budweiser wagon pulled by the famous Clydesdales. Following the parade, in a formal opening ceremony, Fair Director Mel Willard introduced the members of the State Fair Commission. Governor Holden spoke, reminding visitors on the importance of the fair to Missouri agriculture and the importance of agriculture to Missouri. President William Howard Taft, portrayed by Mike Leffler, and former President Harry Truman, portrayed by Niel Johnson, reviewed their visits to the fair.

Other tributes to the past included the Missouri Frontier Exhibit commem-

Facing page, clockwise from top: Signs trace the history of Thomas Bast's buildings. (R. C.)

Ride inspector Paul Zeller checks each ride daily for safety in order to prevent a recurrence of the injuries caused in 1991 when the Gravitron collapsed. (S. B./S. D.)

FFA members join dignitaries in dedicating their new building. (M. S. F.)

Sedalia National Guard members proudly carry the United States and the flag in the state fair parade. (R. C.)

Harry S Truman helped open the one hundredth fair. (S. B./S. D.)

orating Missouri's role in the opening of the West. A one-half scale replica of the keelboat used by Lewis and Clark on their expedition up the Missouri River and a dugout canoe typical of those used in early Missouri highlighted the display. Furs from a variety of Missouri animals reinforced the importance of the fur trade. Frontier storyteller Jim "Two Crows" Wallen, performing in the MoAg Theater, recounted Native American stories explaining why skunks have stripes and why buzzards are bald. In the Agriculture Building, visitors could test their knowledge of old-time implements in a "What Is It?" display of ice tongs, curling irons, hoof trimmers, and nose ringers. At the grandstand, cyclists raced 1870s-vintage and reproduction high-wheel bicycles, comparable to those raced at Association Park when it hosted a fair. One event attempted to answer a question at least as old as the state: how does one pronounce *Missouri*? Should the final syllable be "ree" or "rah"? Secretary of State Matt Blunt polled visitors to find the preferred pronunciation; he discovered state fair visitors are about equally divided between the "Missou-rees" and the "Missou-rahs." Grand opera and agriculture, features of early fairs, came together in 2002 during the Missouri State Fair Queen contest. Fifty-five young women competed for the title, college scholarship, and a silver-plated cup, a replica of the silver cups given to winners at early fairs. Kelley Rohlfing, a recent graduate of University of Missouri's Agriculture Education program and a former 4-H and FFA member, sang "Time to Say Goodbye." Rohlfing reigns as the 2002 Missouri State Fair Queen, telling both fair participants and fair visitors about the wonders of the state and the state fair.

Instead of the "snake-eaters" who entertained and the large groundhog put on display at early fairs, this year visitors could see trained tigers, sharks, and parrots, and pat camels and llamas. Education remained a focus of these exhibits, as animal handlers emphasized the threat of extinction facing many species. Bruno Blaszek's tiger Sheba crossed the staging arena ten feet above the ground on a pair of wires. Other tigers jumped and climbed on command. The Live Shark Encounter featured diver Vincent Fox swimming with the sharks in the tent's large tank and describing the habits of the dangerous creatures. Marcus Beatty, dressed as a pirate, showed off his parrots Salsa, Rainbow, Eco, and Maxwell. The Exotic Animal Petting Zoo allowed children and grownups to engage in personal encounters with camels, zebras, ostriches, and other interesting beasts. The llama show, with prizes for

Old implements are on display in the Agriculture Building. (M. S. F.)

the best llamas at halter and for the best-costumed llamas, indicates an interesting change in Missouri agriculture—the adoption of exotic animals by Missouri farmers.

Vaudeville may not play the stage anymore, but it is alive at the state fair. Ventriloquist James Wedgwood made voices come from nowhere, the one-armed juggler kept balls aloft, legendary aerial artists the Flying Wallendas created pyramids on the tightrope, the a capella quartet the Standards sang, the Recycle Cycle Man blew bubbles from his car made of cast-offs, the Ozark Mountain Cloggers danced, and Buttons the Clown entertained children. Bartholomule, the Missouri State Fair mascot, wore his tuxedo and greeted guests. Contemporary music proved a hit, as both the Kenny Chesney and Toby Keith concerts drew record crowds. Other performers—REO Speedwagon, Styx, Aaron Carter, Brad Paisley, Ted Nugent, Grand Funk Railroad, Sara Evans, Lonestar, the Bellamy Brothers, Lee Greenwood—delighted audiences.

The Wall of Fame highlights FFA achievement. (M. S. F.)

One theme coming from the brainstorming session—the continuing emphasis on speed and power—demonstrated its vitality in the races. The many horse shows, the rodeos, and two days of harness racing reminded racing buffs that the horseless carriage has not replaced the horse completely. The Missouri State Championship races, held on the final night of the fair, identified the best of five classes of drivers and cars—the dirt track late models, modified, street stocks, e-sprints, and late model/super stock cars. This year's race featured one driver, Shorty Acker, who participated in the first state championship at the fair in 1956. Draft horses and tractors both vied for recognition as the most powerful. An antique tractor show brought happy memories to some and sighs of relief to those who prefer larger tractors with air-conditioned cabs. A new event this year showed another kind of speed and strength. A fire engine pull pitted teams of men against one another to see which group of ten men could pull a 47,000 pound fire truck twenty feet in the shortest amount of time. City and state officials comprised a team, but its time of 9.66 seconds revealed that the dignitaries lacked the weight and strength of the winning teams, which boasted times of 6.57 to 7.21 seconds.

Kelley Rohlfing begins her reign as 2002 state fair queen. (S. B./S. D.)

The sale of champions broke all records this year, as the animals raised by young Missourians brought high prices. The grand champion steer brought $15,500, and the grand champion barrow brought $10,500. Smaller stock also brought high prices; the grand champion pen of rabbits sold for $2,200, and the pen of chickens sold for $3,600. At the Ham Breakfast, commemorative posters sold for $2,200, and a belt buckle, donated to the State Fair Museum, brought $1,400. The grand championship ham sold for $7,000. Premiums hit a record amount his year, as over $450,000 was given in prize money to winners of various contests.

Over 360,500 people attended the 2002 Missouri Sate Fair, confirming that the state fair, while a century old institution, is still a vital part of Missouri. As Americans face a frightening future, there is a tendency to look to the past for our security, to seek a reaffirmation of traditional values and ideals. We try to find, in memories of times we see as simpler, though the historians tell us they were neither easier nor safer, a confirmation that all is right with our

Llamas seek help from the Information Booth. (M. S. F.)

world. In the haze of memory, we see a time and place where everything seemed better—a time when an excursion to the state fair was a major expedition for the family, who could go home and share with their neighbors stories of the sights they had seen and the wonders they had beheld. We carry with us memories of being a wide-eyed kid watching the diving mules, of feeling rich when cotton candy cost a dime, of wandering the Midway, of working at the fair to earn money to pay for a semester's tuition or to buy school clothes, of sneaking into the side show or the Pink Garter Revue, of being suspended upside down on the Kamikaze, of playing a piano solo before going to "do the fair." And as we carry these memories, we know another generation of Missourians are building their own memories of animals raised, of prizes garnered, of corn dogs consumed, of stuffed animals won.

Kids admire the Recycle
Cycle. (M. S. F.)

Bartholomule speaks to
young fairgoers. (M. S. F.)

"Celebrate 100 Fairs of Fun"

Centennial Missouri State Fair   August 8-18, Sedalia

A commemorative postcard celebrates the changes in racing. (M. S. F.)

Right: The prices go up and up at the sale of champions. (S. B./S. D.)

The raisers of champion stock smile at having won championships and at selling their animals for record prices. (M. S. F.)

Peter Schauer
wearing a
fiftieth fair tie
celebrates the
one hun-
dredth fair.
(M. S. F.)

Even older kids like to dance with Smokey Bear. (M. S. F.)

Carnival rides delight
and frighten. (M. S. F.)

A sleepy fairgoer naps. (M. S. F.)

Timeline

1899—Missouri General Assembly authorizes the state fair and selects Sedalia as its site.

1901—Missouri General Assembly appropriates $50,000 for construction of buildings.

1901—First Missouri State Fair held, September 9–13, with 25,246 attending.

1903—Varied Industries, Commercial, and Missouri Heritage Buildings erected.

1906—Coliseum built.

1907—Twenty-four acres purchased for use as a campground.

1909—Wright Brothers demonstrate their flying machine.

1910—Womans Building opens

1911—President William Howard Taft visits the fair.

1921—Missouri celebrates 100 years of statehood at the fair.

1925—The Missouri State Fair celebrates its twenty-fifth fair.

1927—Administration Building is dedicated.

1933—The State Fair Board is eliminated and fair management comes under the Department of Agriculture and the Fair Director.

1935, 1936—Miss Missouri Pageant held at the fair.

1943, 1944—Fair canceled because of World War II.

1946—Fair buys additional forty acres.

1950—Merci Train car placed on fairgrounds.

1951—Sally Rand makes her first appearance at the fair. She returns to the fair from 1952 through 1957.

1952—Fair celebrates its fiftieth fair. A tornado damages several buildings and demolishes the Midway, but the fair is back in operation the next evening.

1955—Former President Harry Truman visits the fair.

1958—Missouri State Fair Queen contest begins.

1961—The Agriculture Building, the first climate controlled building on the fairgrounds, opens.

1968—Governor Warren Hearnes dedicates a new grandstand.

1972—The fair builds a shelter house on the campgrounds.

1974—The Ozark Music Festival in July brings hordes of young people to the fairgrounds.

1984—President Ronald Reagan visits the fair.

1988—Exhibition Center is built, and named in 1994 for Senator Jim Mathewson.

1994—Renovations are begun to enable fair buildings to comply with the Americans with Disabilities Act.

1996—General Assembly establishes the State Fair Commission to oversee the fair.

1999—General Assembly appropriates money to fund completion of the Master Plan for fairgrounds renovation.

2002—Missouri State Fair celebrates One Hundred Fairs of Fun.

WORKS CONSULTED

An Appeal from the Industrial Interests for the Creditable Equipment of the State Fair Grounds. 1901. State Historical Society of Missouri. Columbia, MO.

Boan, Jim. "Missouri State Fair." unpublished document, 1947.

Chillicothe Star. 27 April–8 June 1899, passim.

Christensen, Lawrence O. and Gary R. Kremer. *A History of Missouri: Volume IV, 1875 to 1919*. Columbia: University of Missouri, 1997.

Claycomb, William B. *Pettis County, Missouri: A Pictoral History*. Virginia Beach: Donning, 1998.

_____. "Missouri Architects and Builders: Thomas Bast, 1863–1933." *Preservation Issues* 4.2 (1994): 4–5.

"A Clover is Born . . . History of 4-H." 2 Feb. 2000. 29 December 2001.
< http://www.reeusda.gov/4h/4h history.htm >

Columbia Herald. 12 May–9 June 1899, passim.

Conaway, Jim. "State Fair Keeps Grip on Racing." *Kansas City Star*, 25 August 1994.

Conservation Commission of the State of Missouri. *The History of the Conservation Movement in Missouri*. 2nd ed. Jefferson City, MO: 1990.

Curtis, Susan. *Dancing to a Black Man's Tune*. Columbia: University of Missouri, 1994.

Curran, Pat. "Norman J. Colman." *Dictionary of Missouri Biography*. Ed. Lawrence Christensen, et al. Columbia: University of Missouri, 1999.

Dains, Mary. "The Missouri State Fair: The Struggle to Begin." *Missouri Historical Review*, 73 (1978):23–53.

Ensminger, M.E. *The Complete Encyclopedia of Horses*. New York: Barnes, 1977.

"Evolution of the Stock Car." NASCAR.com website. 6 February 2002. 9 pp. 25 June 2002.
< http://www.nascar.com/2002/kyn/history/evolution/02/06/stockcar/index.html > .

"4-H Through the Decades." unpublished essay, 2002.

"History of 4-H at the Missouri State Fair." unpublished essay, n.d.

"The History of the National FFA Organization." FFA History. 29 December 2001.
< http://www.lenoir-hs.loudon.k12.tn.us/lcpage/clubs/ffa/history.htm >

Gaskell, Richard. *The Missouri State Fair: Images of a Midwestern Tradition*. Columbia: University of Missouri, 2000.

Goldsack, Bob. *Cetlin and Wilson: One of America's Great Railroad Carnivals*. Nashua, NH: Midway Museum Publications, 1987.

Gordon, John Steele. "The Chicken Story." *American Heritage*. Sept. 1996: 52–67.

Grace, Karen. "The Fair Women of Missouri." *Preservation Issues*, 4.2 (1994): 6.

Guitar, Sarah. Western Missouri Manuscript Collection 3563, folder 63. Sarah Guitar Papers. University of Missouri, Columbia.

Hagan, Raymond. Personal interview. 18 May 2002.

Herrington, Thomas W. "Norman J. Colman." *Prominent Men and Women of the Day*. 1888. 3 March 2002.
< http://www.cyberschool.k12.or.us/ ~ layton/biogrraphies/c/normanjcolman/normanjcolman/html > .

"History of Midget Racing." MidgetRacing.com website. 7 July 2002.
< http://www.midgetracing.com/history.httm > .

Homan, Anneliese. "The 1974 Ozark Music Festival: 'No Hassles Guaranteed'" unpublished essay, 2000.

Hooker, Roger D. *History of the 509th Bomb Wing, Whiteman Air Force Base*. unpublished manuscript, 1993.

Jennings, Ron. "100 Years: Fair Celebrated State's Century Mark." *Sedalia Democrat* undated clipping in Missouri State Fair Collection, Sedalia Public Library.

_____. "Past Revisited Through Exhibit." *Sedalia Democrat*, 24 August 1976.

_____. "Sedalians Remember When Fair Was Born." *Sedalia Democrat*, 14 August 1977.

Kennedy, Charles A. "When Cairo Met Main Street: Little Egypt, Salome Dancers, and the World's Fairs of 1893 and 1904." *Music and Culture in America*. ed. Michael Saffle. New York: Garland, 1998.

Kern, James P. Western Missouri Manuscript Collection 2700, folders 12201–12011. James P. Kerns Papers. University of Missouri, Columbia.

Kirkendall, Richard S. *A History of Missouri: Volume V, 1919 to 1953*. Columbia: University of Missouri Press, 1986.

Kniffen, Fred. "The American Agricultural Fair: The Pattern." *Annals of the American Association of Geographers* 1 (1949): 264–282.

_____. "The American Agricultural Fair: Time and Place." *Annals of the American Association of Geographers* 3 (1951): 42–57.

Lang, Hazel. "The Missouri State Fair." *Life in Pettis County*. 787–794.

Lloyd, Nelson. "The County Fair." *Scribner's Magazine* August 1903:129–147.

Mares, Fred. "Missouri State Fair Gambles on New Sorts of Fun." *Kansas City Times*, 15 August 1986.

Marti, Donald. *Historical Directory of American Agricultural Fairs*. Westport, CN: Greenwood Press, 1986.

Maserang, Roger. "A Missouri Classic: The Missouri State Fairgrounds National Register Historic District." Ed. Karen Grace. Jefferson City: Department of Natural Resources, n.d.

_____. The Missouri State Fairgrounds Historic District. Nomination to the National Register of Historic Places.

Mexico Intelligencer. 30 May–6 June 1899, passim.

Meyer, Duane G. "Beck, Helen Gould (Sally Rand), *The Dictionary of Missouri Biography*. ed. Lawrence Christensen, et. al. Columbia: University of Missouri Press, 1999.

Missouri Council of Defense. Western Missouri Manuscript Collection, 11, folder 1191. Missouri Council of Defense Records. University of Missouri, Columbia.

_____. Western Missouri Manuscript Collection, 2797, folders 106, 637–641. Missouri Council of Defense Records. University of Missouri, Columbia,

Missouri Department of Conservation. "Wildlife Pavilion: A Conservation Message at the State Fair." unpublished document, n.d.

"Missouri Farm Facts, Farm Income and Prices Summary." Missouri Agricultural Statistics Service Website. 6 June 2002. < http://agebb.missouri.edu/mass/farmfact/farmfact.htm > .

Missouri Ruralist, 8 August 1964, passim.

Missouri State Fair Collection, Missouri State Archives, Jefferson City, MO.

Missouri State Fair Collection. Sedalia Public Library. Sedalia, MO.
.
Missouri State Fair Board of Directors. "Minutes." 1899-1933, passim. State Historical Society of Missouri. Columbia.

Missouri State Fair: Premium List. 1901–2002, passim.

Missouri State Fair: Program Guide. 1901–2002, passim.

Missouri State Fair: Report of the Disinterested Comment of Agricultural and Livestock Journals. Sedalia, MO: Hodges, 1901.

Missouri State Highway Patrol: 70th Anniversary, 1931–2001. n.p.: n.p., 2001.

Missouri Statesman. 14 April 1899.

Moberly Evening Democrat. 18 April–8 June 1899, passim.

Neely, Wayne Campbell. *The Agricultural Fair*. 1935. New York: AMS Press, 1967.

Nelson, Derek. *The American State Fair*. Osceola, WI: MBI, 1999.

Nichols, Kent. "Keep the Past, but Move to the Future." *Sedalia Democrat*, 14 March 1996.

Park, Guy B. Western Missouri Manuscript Collection 8, folders 1390-1393. Guy B. Park Papers. University of Missouri, Columbia.

Patterson, Connie S. "Growing Up at the Fair." *Missouri Resources* 15.2 (Summer 1998): 5.

Perrin, Noel. "The Old State Fair, Still What Used to Be—Even More." *Smithsonian* Sept. 1985: 96–109

"Progress Made Visible: American World's Fairs and Expositions." University of Delaware Library Special Collections Department Website. 4 September 2001. < http://www.lib.udel.edu/ud/sppec/exhibits/fairs/column.htm > .

Roberson, Ernie. "Gus Schrader—King of the Outlaw Dirt Racers." ArtLatexRacing Report.com website. 24 June 2002.
< http://www.artlatexracingreport.com/Story_Gus_Schrader.html > .

Ruhl, Arthur. "At the County Fair." *Collier's, The National Weekly*, 16 August 1913: 20–21, 34.

Sandy, Adam. "Roller Coaster History." Ultimate Roller Coaster Website. 2001. 15 July 2002. < http://www.ultimaterollercoaster.com/coasters/history > .

Sedalia Capital. 3 May 1901–23 August 1902, passim.

Sedalia Democrat. 1899–2002, passim.

Sedalia, Missouri City Directory. 1898–1899, passim.

Sedalia, Missouri City Directory. 1900–1901, passim.

Sedalia, Missouri: 100 Years in Pictures. Sedalia: Sedalia Area Chamber of Commerce, 1960.

Sedalia, Missouri: The Commercial, Industrial and Educational Metropolis of Central Missouri. Sedalia, Missouri, Sedalia Evening Sentinel, 1904.

Sedalia, Missouri. The First One Hundred Years. Sedalia, MO: Hurlbut Printing, 1960.

Sedalia Sentinnel. 1 May 1899–22 September 1903, passim.

Sisemore, Rhonda Chalfant. *An Illustrated History of Sedalia and Pettis County*. ed. F. Douglas Kneibert. Jostens, 1990.

Smith, J.D. *Souvenir: Missouri State Fair, 1917*. Sedalia, MO: Sedalia Printing, 1917.

"Soybeans," *Encyclopedia Americana*, 1904.

State Board of Agriculture. *Report*. Jefferson City, MO: 1873, 1893, 1901, 1902, 1912, 1916.

Stephen, Les and Ward Sullivan, Jr. *Cattle Breeds Index*. Hays, KS: Research Communications, 1976.

Stephens, Frank F. *History of the University of Missouri*. Columbia: University of Missouri Press, 1962.

Stout, Laurie and Mark Schreiber. *Somewhere in Time: A 160 Year History of Missouri Corrections*. Jefferson City: Missouri Department of Corrections, 1991.

Tevis, Cheryl. "No Farewells to Ag Fairs." *Successful Farming*, August 2000, 3 November 2001.
< http://www.finddarticles.com/cf_sccfrm/m1204/9_98/64830743/print.jhtml > .

Thompson Museum Consulting. "Farming, Family, and Fun: 100 Years at the Missouri State Fair." Exhibit Conceptual Plan. 19 September 2001.

"Tracking the Past to the Present." *Sedalia Democrat*, 22 July 1994: SS 1–7.

"The History of Pulling." Tractor Pulling History Website. 30 June 2002.
< http://www/geocities.com/Hoollywood/Set/9671/Pulling/history.html > .

United States Census, 1910, Agriculture. 13 June 2002.
< http://fisher.lib.vigrinia.edu/cgi-local/censusbin/census/cen.pl >

United States Department of Agriculture. Commodity Reports. 6 June 2002.
< http://www.nass.usda.gov:81/ipedb/report.htm >

University of Missouri Extension Service Bulletin 701. Columbia: University of Missouri,
1952.

Utz, Bill. Interview with Doug Jones. *Sedalia Democrat*. 22 July 1994: SS2, 5.

Viles, Jonas. *The University of Missouri, 1839–1939*. Columbia: University of Missouri,
1939.

Williams, Walter, and Floyd Calvin Shoemaker. *Missouri, Mother of the West.* Vol. II,
Chicago: American Historical Society, 1930.

Yost, Don and Connie S. Patterson. "Meet us at the Fair." *Missouri Resources* 15.2
(Summer 1998): 2-5.

ABOUT THE AUTHOR

Local historian Rhonda Chalfant is a member of the Language and Mass Communication Department at State Fair Community College in Sedalia where she teaches composition and literature classes. She holds an M.A. in English from Central Missouri State University and an A.b.D. in History from University of Missouri, Columbia. Active in the Pettis County Historical Society and local preservation efforts, she seeks to combine her love of "neat old stuff" with her studies of nineteenth century social history. Her goal is to write thoughtful, academic history for the popular market. She writes a weekly history column in the *Sedalia Democrat*, and makes regular presentations on women's, ethnic, and class history. She and her many cats live in Sedalia.